T3-BNI-686

HYPOTHESIS-TESTING BEHAVIOUR

Hypothesis-testing behaviour

Fenna H. Poletiek
University of Leiden, The Netherlands

First published 2001 by Psychology Press Ltd
27 Church Road, Hove, East Sussex, BN3 2FA

www.psypress.co.uk

Simultaneously published in the USA and Canada
by Taylor & Francis Inc
325 Chestnut Street, Suite 800, Philadelphia, PA 19106

Psychology Press is part of the Taylor & Francis Group

British Library Cataloguing in Publication Data
A catalogue record for this book is available from the British Library

Library of Congress Cataloging-in-Publication Data
Poletiek, Fenna.
Hypothesis-testing behaviour / Fenna H. Poletiek.
p. cm. — (Essays in cognitive psychology, ISSN 0959-4779)
Includes bibliographical references (p.) and indexes.
ISBN 1-84169-159-3
1. Hypothesis. 2. Psychology—Research. 3. Psychology, Experimental.
I. Title. II. Series.

BF76.5 .P59 2001
153.4'3—dc21

00-059070

ISBN 1-841-69159-3

Cover design by Code 5 Design Associates Ltd
Typeset in Times by Mayhew Typesetting, Rhayader, Powys
Printed and bound in the UK by Biddles Ltd, www.biddles.co.uk

For Joske, Arianne, and Lys

Contents

Foreword

How can people be sure that their ideas reflect the state of nature, which is what hypothesis testing research is about? And how can hypothesis-testing researchers be sure that their ideas reflect how people actually test their ideas? These questions are what the present essay is about. Peter Wason found two brilliant ways—two experimental tasks—to put his subjects' testing behaviour to the test. In these tasks his subjects displayed a pervasive confirmation bias. This was at odds with what logicians and philosophers—the most prominent being Karl Popper—had prescribed to be rational for the human hypothesis tester. Both the experiments and this finding resonate in all the work on hypothesis testing since the early 1960s. But most of this follow-up work is dissonant as well. Confirmation bias is sometimes argued not to be the irrational way of testing it was supposed to be, or it disappeared when small adaptations of the tasks were made, or it was proposed to be an artefact of the experimental situation, and so on. All researchers have their own special relationship with the work of Peter Wason, and the present essay is no exception.

Peter Wason was intending to write this Foreword. In his letter of 4 May 1999 he announced: "My theme will be how much the investigator will miss if he delegates the running of the experiment to an assistant, as is universal in the USA! What do you think?" Well, I can imagine a witty, sharp, and joyous argument leading to the rehabilitation of confirmation bias. It might be the merciful confirmations of his ideas the experimenter would miss! But we don't know. In his letter of 14 September 1999, he sadly had to decline the invitation to write the Foreword because of ill health.

I am indebted to him for launching hypothesis-testing research so vigorously that it remains a vital source of inspiration more than 30 years later, and it probably will be for the future generation of cognitive psychologists who want to find out how people test their ideas. Also, I thank Jonathan Evans, Nick Chater, and Ken Manktelow for their helpful comments on earlier versions of this book. Theo Vosse is thanked for his mathematical assistance. Finally, Ken Manktelow is thanked again, together with George Hall for taking care of my English at several stages of the manuscript.

Amsterdam, May 13th, 2000

Introduction

Assuming that the mind is meant to help us cope with our environment, it could be stated that its major task is that of hypothesis testing. Along this line of thought, many functions of the mind may be described as instances of hypothesis testing. Visual perception, then, is continuously testing hypotheses about the visible world by means of looking. Learning a language, for instance, consists of uttering sounds and eventually tentative sentences, and testing whether they are "correct" by observing a listener's reaction. Problem solving can be seen as trying out hypothetical solutions by observing their effect in the world. Many higher mental processes are obvious variations of hypothesis testing. In sum, hypothesis testing comprises comparing internal thoughts to external facts in order to interact with the world.

The study of Bruner, Goodnow, and Austin (1956) was one of the first fundamental empirical investigations of our cognitive commerce with the environment. Their focus was on how people place observations in theoretical categories, by means of hypothesis generation and testing. Bruner et al. (1956) also consider this cognitive "activity" as ubiquitous, and as a prerequisite of any other cognitive task performance. Indeed, in order to process and produce any new information, old information must have been categorised in abstract hypotheses, which in turn have been tested against previous experience. From this point of view, the subject is extremely broad and complex. Notwithstanding this complexity, the purpose of this book is to present a concise empirical and theoretical essay. Therefore, a more restricted definition of hypothesis-testing behaviour is proposed.

Hypothesis testing is looked upon as a cognitive activity that can be divided into stages. Initially, the tester has a hypothesis about the state of nature with regard to some subject. Next, he or she chooses or designs a test whose outcome is expected to reveal something about the truth status of the hypothesis. Then the test is performed and the result observed. Finally, the result is interpreted as to its consequences for the original hypothesis. This definition still covers a large range of cognitive activities. Sometimes, research focuses on the early stage of the process: Generating a hypothesis and designing a test. Sometimes, the later stages are also investigated: The way in which the results are interpreted in the light of the hypothesis. The second stage, in particular, is the one we look at in the present study. Thus, unlike Bruner et al., hypothesis generation is not focused on here. It is assumed that the tester has some hypothesis or hypotheses in mind that he or she intends or is asked to evaluate in the light of some evidence. I do not go in detail into how the idea has come to the tester's mind. However, the last stage of interpretation of the test result will also get some attention. Indeed, the estimated impact of the test result on a hypothesis can be an important reason for choosing a test in the first place.

Another limiting assumption is that the hypothesis tester looks for *new* evidence. Thus, the information sought is unknown, and not available, at the moment that people choose or design a test whose purpose is to provide that information. This aspect makes hypothesis-testing behaviour more complex than it might at first seem. Indeed, the issue at stake is not: "Which information or evidence from some body of evidence do people select or prefer in order to evaluate their hypotheses?" It is rather: "Which experiment is performed, which questions are asked, which population is observed, which person is consulted, or what is uncovered, in order to get the information that will be used to evaluate the hypothesis?" Take, for example, a judge who must make a decision about the guilt of an accused. The decision about which investigations the judge will perform in order to gather evidence (or counter-evidence) about the accused's guilt is an example of testing behaviour. If, however, various pieces of evidence are already known to the judge, and he or she selects some of these to underpin the hypothesis of guilt, then he or she is involved in evidence interpretation. The latter behaviour is characteristic of the final stages of hypothesis testing: The interpretation of test results in the light of the prior hypothesis. However, the interpretation and perception of available evidence is a different behaviour from test choice (the central stage), even if the two stages are not always strictly distinguishable in practice.

The following chapters elucidate further why the very definition of hypothesis testing makes its study so very complex. For example, a difficulty lies in defining *strategies* of testing behaviour. How does one grasp and demarcate the strategies people use to look for the unknown? What should be understood by *confirmatory-testing* behaviour, for example? What kind of tests (questions, experiments) might reveal that the hypothesis tester is biased towards finding confirming evidence, when we also assume that she cannot know where to find it, since it is new evidence? What is "confirmatory testing" in practice? Clearly, a confirmatory strategy will consist of selecting only confirmatory evidence when several pieces of ambiguous evidence are already available—but this is an interpretation and not a testing strategy. Essentially, a confirmatory-*testing* strategy is an awkward concept. Indeed, the very intention to test some idea against some evidence means that one is interested in knowing "the truth" about it (Poletiek, 1996). Why look for new evidence about a belief, undergoing the risk of falsification, if we want to stick to the original belief anyway? And *how* could we possibly manipulate a test beforehand, in such a way as to control its result, given that the data we look for are unknown? These intuitions will be worked out in more depth when the theories and experiments on hypothesis-testing behaviour are discussed. They form a major motivation for conducting the present study.

Why, how, how long, and by whom has hypothesis-testing behaviour been studied by psychologists in the past? After the Bruner et al. (1956) study, hypothesis testing was explicitly studied for the first time in psychology by Wason (1960). Before this date, hypothesis testing typically belonged to the realm of philosophy of science. There is a huge amount of philosophical literature about how hypotheses are tested in science, and about how they should be tested. The most well-known work on this topic is that of the philosopher Popper (1959/1974). His theory of falsificationism has had an enormous influence in almost all disciplines of science. It proposes that theories in science should be exposed, as much as possible, to empirical falsification. This is the best way to obtain growth in knowledge. Besides, Popper (1959/1974, p. 18) claimed that this not only applies to science, but also to everyday human reasoning and learning about the environment:

> . . . I agree that scientific knowledge is merely a development of ordinary knowledge or common sense knowledge . . .

Earlier, the psychologist Kelly (1955) had already compared human knowledge acquisition to scientific reasoning. He proposed that humans are intuitive scientists proceeding in basically the same way as scientists do, but less systematically and explicitly.

Thus, although hypothesis testing had long been a problem of philosophers of science only, the global idea that scientific knowledge acquisition could serve as a model for general human knowledge acquisition did already exist. Peter Wason was the first to explicitly use the philosophical falsificationist model of hypothesis testing to explain psychological data. He was one of the students in London at the end of the 1950s who were fascinated by the new theory of falsificationism advanced by Popper. The Viennese philosopher had been appointed in London at the time the English version of his *Logik der Forschung* (Popper, 1935), in which he had worked out falsificationism, was published (*Logic of Scientific Discovery*, Popper, 1959/1974). Wason applied the new principle to human reasoning by means of very elegant and shrewd experiments. He gave the impulse for a stream of theorising and experimentation about human hypothesis testing which has not ceased right up until the present day (Newstead & Evans, 1995). Interestingly, Wason's very first experiments are still the most replicated and analysed in the hypothesis-testing research of recent years.

How can the success of Wason's experimental programme be explained? At least two factors have played a role. First and foremost, the experiments themselves are fascinating due to the contrast between their apparent simplicity at first sight, and the huge complexity in which researchers find themselves when attempting to explain responses to them. This complexity will become clear when I discuss the theorising on Wason's experiments in Chapters 3 and 4. In addition, it was found that people did not conform to the Popperian falsification norm when responding. This *cognitive bias* is a robust finding replicated many times. Cognitive biases have always attracted a lot of attention in the scientific community. The impressive research of Kahneman, Slovic, and Tversky (1982) is an example of this. In this research, similar to the earlier Wason experiments, participants are presented with a problem that can be solved in a formal (logical, mathematical, or statistical) way. The participants' responses are observed and compared to the formal solution. A non-fit is interpreted as a cognitive bias in the human reasoner (Evans, 1989).

However, shortly after the emergence of the "cognitive biases" paradigm, critics felt that something was wrong with bias research. Essentially, its prescriptive character and its lack of pragmatism in explaining subjects' behaviour were most problematic. But in spite of the nuances which have been advanced, "confirmation bias" lives its own life in the scientific community as well as outside it (Evans & Over, 1996a). Cognitive biases like these are good survivors in psychology. In this book, this debate in hypothesis-testing research will be discussed. Because of the central role of Wason's work, and the fact that all hypothesis-testing research is related directly or indirectly to it, two chapters are dedicated to his two most famous problems, the rule discovery task and the selection experiment.

The selection task is the most used vehicle for hypothesis-testing research. This simple task, originally a pure conditional reasoning task, has evolved to become the standard paradigm in research on a huge variety of behaviours related to information search and processing. However, in its evolution, the selection task has more and more become questioned as to what exactly it stands for. The statistical models of the selection task, emerging in the beginning of the 1990s, are greatly responsible for the extension of the selection task from a logical task to a general hypothesis-testing task. Indeed, the new statistical views explain effects of prior beliefs, uncertainties, pragmatic goals, etc., on the strategies people use to look for evidence. Before this development, proposition logic was the standard for solving the task. This may be an adequate model for conditional reasoning (although that idea is itself debatable), but it is a quite implausible model for describing general and everyday processes of hypothesis testing. The evolution of the selection task demands thorough discussion, especially the new view of it, which has contributed substantially to our knowledge of hypothesis testing.

This book is predominantly theoretical. The underlying policy was to go back to the theories on hypothesis testing from disciplines like philosophy, logic, and statistics and to discuss the empirical work against this background, rather than to start from the empirical data. I believed it to be more useful to take up a theoretical and integrating study, although the present contribution by no means claims to have spoken the final word on the matter. Some old puzzles and contradictions in the theorising on hypothesis testing have been put together in the theoretical section, such as the contrast between confirmation and falsification, for example, and looked at from a new perspective. Importantly, much work on hypothesis testing is not discussed here. This is a consequence of the intention to deal with the major developments in depth rather than to give a comprehensive survey.

I start with theories of hypothesis testing. There is a large literature on how hypotheses should be subjected to empirical testing. These theories originate especially in the philosophy of scientific reasoning. The psychology of hypothesis testing relies heavily on these theories when modelling human "everyday" hypothesis testing. However, very few psychological studies focus on this theoretical background itself, and how it should be translated to empirical work. The theories are generally merely mentioned or assumed and translated into experimental set-ups. The advancement of psychological investigation into hypothesis-testing behaviour can benefit greatly from a thorough understanding of what the philosophy and mathematics of testing basically stand for. It is a major purpose of the present book to contribute to this interaction.

Chapter 1 discusses philosophical theories of scientific hypothesis testing. The focus is on the way they represent "good testing", their psychological implications and their—striking—mutual similarities. Chapter 2 deals with formal theories of testing. These statistical theories are also mostly oriented towards scientific hypothesis testing. The same focus is applied as that used for philosophical theory: I look at the coherence between different theories and, in addition, compare them with the philosophy of testing. From

Chapter 3 on, experimental research into hypothesis testing is discussed. This work is divided into three categories, according to the methodology used. The first and second categories refer to the two standard tasks by Wason. These are still the paradigms that have produced by far the most knowledge on hypothesis testing, especially the selection task, as was argued above. Chapter 3 deals with the rule discovery task and its spin-offs. Many replications and analyses are presented which, in turn, are analysed against the background of the theories of testing presented in the foregoing chapters.

Chapter 4 is about the selection task. A short survey of "old" findings and interpretations of this intriguing experimental task is followed by a thorough discussion of some major recent statistical analyses. Here again, I repeatedly step back to the theories from the first two chapters. It will be shown how the probabilistic modelling of a non-probabilistic task has produced valuable knowledge about hypothesis testing, and also how it runs into problems due to this paradox. Chapter 5 deals with probabilistic models of hypothesis-testing behaviour in probabilistic tasks. These experiments rely greatly on the statistical theories of testing discussed in Chapter 3. I also introduce the probability-value model of hypothesis testing. This model is consistent with the philosophy of testing and integrates well with the new analyses of testing behaviour in the selection task, the rule discovery task, and the probabilistic experiments. It can solve persistent problems with regard to the confirmation-falsification paradox. Most importantly, it is quite intuitive. The book rounds off with Chapter 6 in which the probability value model is referred to again. First, the rationality question is addressed. Second, the probability value model is set out as a general perspective for hypothesis-testing behaviour. Also, the implications of the probability value model for human rationality are investigated. This approach unifies pragmatic considerations with formal rationality. Third, I consider the implications of the present study for both everyday and professional hypothesis testing, and sketch some new lines of investigation in the study of hypothesis-testing behaviour.

CHAPTER ONE

Theories of testing in the philosophy of science

INTRODUCTION

This chapter provides a survey of theories of testing in the philosophy of science. These theories are frequently used as points of departure in psychological research on testing behaviour, with the purpose of describing and explaining the conduct of human testers. The central questions governing the discussion of the theories are: What is regarded as a rational manner of searching and using evidence when judging a theory or hypothesis, and what is seen as irrational or biased? The crucial concepts borrowed from the philosophy of science by cognitive psychology in order to study human testing behaviour are those of "confirmation" and "falsification". These concepts play a central role in the philosophies of the science of logical positivism (Carnap, 1936–1937) and Popper's critical rationalism (1959/1974, 1963/1978). Most psychologists have adopted the principle of falsification rather than that of confirmation as a standard for human hypothesis testing.

Borrowing this standard from the philosophy of science, psychologists assume that the scientific situation, for which these standards have been developed, is relevant to the general situation of humans trying to acquire knowledge about their environment. On the other hand, philosophers (e.g., Popper, 1963/1978) have emphasised that scientific knowledge acquisition is a systematised version of knowledge acquisition in ordinary life. Because of the great impact of the philosophy of science, and especially of the concepts of confirmation and falsification in the psychological study of human testing behaviour, the background of these concepts will be thoroughly discussed. What makes, in the authors' views, a principle into an adequate guideline to test scientific theories, and why is falsification preferred to confirmation as a normative principle in cognitive psychology? How do confirmation and falsification relate to one another? Is the contrast between the two principles, which is much emphasised in the psychology of testing, justified? Criticism of the two standard theories of science is also discussed, since some authors in psychology make use of these critical philosophies in their study of human hypothesis testing.

The theories of science are dealt with in three sections, moving from philosophical views about theory and evidence to more concrete theories concerning the way tests should be performed. The first section introduces the principles of confirmation and falsification, starting with the classical theories of the logical positivists and Popper, and reviewing criticisms of both. The next section deals with these principles at the methodological level of testing individual hypotheses. The term "methodological level" is used to indicate the concrete level at which the philosophical principles are applied in science. It also refers to the formal elaboration of the two philosophies. This section ends with an integration of the two principles. Whereas, at a philosophical level, the contrasts between the two key theories of science will be emphasised, at a methodological level, their concordance will be demonstrated. Subsequently, some remarks are made on the implications of the previous philosophical analyses for the psychological study of human testing behaviour. The chapter finishes with a short concluding section.

PHILOSOPHICAL PERSPECTIVES ON THEORY AND EVIDENCE

One of the major topics in the philosophy of science is the relationship between theories and empirical data. The work of Popper and Carnap deals with how this relation should be justified in scientific work. Other philosophies do not aim at a rational justification but merely describe how, in science, theories are related to data. In this review, logical positivism and falsificationism will be presented, followed by a survey of the critical reactions generated by these main theories.

Logical positivism

Logical positivism is the philosophy of science developed by the members of the Viennese Circle in the first decades of the 20th century. The most important motive of these philosophers was to rid science of metaphysics and "meaningless" statements. These meaningless statements were literally statements without significance. In order to distinguish between meaningful and meaningless statements, the logical positivists applied the criterion of "capable of being verified" (Carnap, 1936–1937). Thus, statements only have meaning if they are capable of being verified. How can theoretical statements, or in short, theories, be verified?

The philosophy of logical positivism consists of a "reconstruction" of science. The concept of "reconstruction" means that scientific developments are described by logical positivists as being a set of logical theories that have an empirical foundation and are not merely historical processes. Accordingly, it is a *rational* reconstruction. This reconstruction, later called the "received view", is a logical system consisting of two languages: a theoretical language and an observational language. Theoretical terms can be reduced to terms from the observational language. Terms in the observational language refer directly to perceivable objects, or perceivable qualities of objects. The logical system of positivism can be regarded as a pyramid where the base is a collection of statements concerning perceivable objects and qualities, and where the peak consists of abstract theoretical statements that are linked to the base. The statements at the base (the "protocol sentences") in the observational language are directly verifiable by an observer. But how are the observational statements (which, after all, are mini-theories in themselves) actually verified?

Developing a theory of verification was an important goal for the logical positivists because they wished to ensure that theories could be "anchored" in "true" observations. Carnap twice produced a theory concerning the verification of observational statements. The first theory of verification (Carnap, 1928) posits that the statements at the base (the protocol sentences) have the character of statements about experience. This is the so-called "phenomenalistic" approach to verification. The kernel of this vision is that the truth of the protocol sentences can be established immediately and indisputably. Accordingly, the statement "I have the perception of red" can be immediately confirmed or refuted by each individual.

The weakness of this vision of basic statements is that their verification varies from individual to individual. In other words, verification is indeed indisputable but it is not guaranteed intersubjectively. After all, different people may have different perceptions of the same object. The second theory regards the protocol sentences as statements about the qualities of objects, instead of about visual perceptions (Carnap 1936–1937). These statements are considered to be intersubjectively verifiable. But one is less certain that the statement is true, for instance, whether the colour *really is* red, since the judgement could be incompatible with a subjective perception. Thus, in the second theory, one observes and *decides* whether the statement has been verified or falsified. Consequently, verification here contains an element of convention. These are the two theories produced in the logical positivist framework which aimed at linking theoretical terms to observations.

Eventually, by means of these theories, meaningful (scientific) statements could, in principle, be distinguished from meaningless statements.

Strictly speaking, the verification criterion is not a criterion for testing a theory. That is, it is not a set of rules which prescribes how, in a deductive process, a given theory can be rationally tested. It tells how a theory should be *induced* from observations in the first place, before being tested. Verificationism is mainly concerned with this inductive process. The fact that the theory is induced from verified observations gives it the status of scientific knowledge. Verification, as used in logical positivism, belongs to an inductive view of science. Suppe (1977, p. 15) emphasised the inductive nature of the received view as:

> Thus science proceeds "upwards" from particular facts to theoretical generalisations about phenomena, this upward process proceeding in an essentially Baconian fashion.

Thus, an essential aspect of logical positivism is that it dictates searching, to the greatest extent possible, for verifiable observations which "confirm" a theory. Carnap's confirmation theory will also be discussed when dealing with the consequences of the verification principle at a methodological level. It provides guidelines as to *which* observations (protocol sentences) should be pursued in order to confirm a theory as being a scientifically true theory.

Before doing so, the most important opponent of logical positivism will be introduced: Karl Popper. He substituted the inductive representation favoured by the logical positivists with an explicitly deductive view of science. Popper shifted the focal point from the induction of hypotheses based on observation, to the testing of any given hypothesis.

Popper's critical rationalism

While the logical positivists were attempting to exclude metaphysics from science, Popper placed the accent on the *growth* of scientific knowledge. According to Popper (1963/1978, p. 215), scientific knowledge can grow when one repeatedly endeavours to *refute* theories and to replace them with better ones:

> It is the way of its growth which makes science rational and empirical; the way, that is, in which scientists discriminate between available theories and choose the better one . . .

Popper's philosophy of science is referred to as being "critical" because of the importance of theory *selection*. Theories must always be evaluated critically so that one can improve or replace them. Accordingly, Popper posits the falsification principle instead of the verification principle.

A great deal of Popper's work contains criticism of the verification principle. In this criticism, the verification principle is not merely regarded as a method of evaluating basic statements with respect to their truth or falsehood. It has many more negative consequences (Popper, 1963/1978). The verification principle is regarded as one that leads to stagnation and dogma when pursued by researchers. This criticism by Popper has an explicit psychological nature in addition to the philosophical one, in the sense that verificationism seems to be condemned as an *attitude* (1963/1978, p. 229).

> . . . the rationality of science lies not in its habit of appealing to empirical evidence in support of its dogmas—astrologers do so too—but solely in its critical approach: In an attitude which, of course, involves the critical use, among other arguments, of empirical evidence (especially in refutations).

The psychological and normative tone of this criticism has contributed to the formation of a rich breeding ground for its application in the psychology of reasoning. Popper's theory became prescriptive for most psychological research on test behaviour. Peter Wason, at that time a student in psychology inspired by the theory of Popper, adopted his normative view. Wason published the first psychological experiment concerning human testing of hypotheses. In this publication, he wrote (1960, p. 139):

> The task simulates a miniature scientific problem . . . . The kind of attitude this task demands is that implicit in the formal analysis of scientific procedure proposed by Popper (1959). It consists in a willingness to attempt to falsify hypotheses, and thus to test those intuitive ideas which so often carry the feeling of certitude.

From Popper's and Wason's standpoint, the falsification principle is depicted as an open-minded strategy, in contrast to the verification principle which is a conservative strategy directed towards repeatedly confirming that which is already established. But it is debatable whether this interpretation corresponds to the intentions of the positivists. They too wished to remove metaphysical theories, such as astrology, from science and in this respect certainly did not wish to propagate any conservative attitude.

The philosophical part of Popper's argument against verificationism, as opposed to the psychological part, runs as follows: A universal statement cannot be verified purely logically by a limited number of observations, but it can be falsified. One single contrary example is sufficient to refute the statement, according to the *modus tollens* reasoning. The famous white swans problem can serve as an example. The universal statement "all swans are white" is not definitely verified when one has observed a number of white swans. Even a great many white swans are not enough to make it true.

However, one black swan makes the statement definitely false. Thus, according to Popper, verification fails on grounds of principle because the demands of verifiability can never be satisfied in cases of universal statements. And the main aim of science is to produce true universal statements.

Popper's recommendation to attempt to refute theories thus essentially consists of two kinds of arguments, each of which is common in psychological studies of hypothesis testing: a logical and a psychological (or moral) argument. However, the arguments have quite different impacts. Interestingly, they do not necessarily lead to the same prediction about testing behaviour. For example, the logical argument for falsificationism does not necessarily lead to the psychological advice to attempt to refute one's *own* theory. A researcher may attempt to refute categorically the alternative to the theory in which he or she believes, as happens in "significance testing". The falsification strategy is then pursued because of its logical rewards. However, this is not in accordance with the psychological recommendation to examine one's own theory as critically as possible. The researcher attempts to reject the null hypothesis and, by succeeding, to give plausibility to the alternative hypothesis which he or she believes to be true. But the researcher may also attempt to confirm the alternative theory in line with the psychological recommendation, as Hofstee (1980) shows. This is critical with regard to one's own theory, without being Popperian in the logical sense. Psychologically viewed, the attempt to refute a theory in which one believes is completely different to the attempt to refute a random or alternative theory. This distinction forms the basis of the discussion in psychology about whether people's tendency to confirm is cognitive or motivational. This discussion is taken up further in this book.

One last aspect of Popper's philosophy that deserves attention is the purpose for which the falsification principle is an instrument: the growth of knowledge. Indeed, both in science and everyday reasoning, the ultimate goal of hypothesis testing is to acquire knowledge. How does Popper argue that hypothesis falsification leads to more readily accepted hypotheses, and therefore to the growth of knowledge? To discuss this, a short overview of Popper's idea of induction is necessary. Popper is opposed to induction. In contrast to the logical positivists, Popper does not place any demands on the way a theory originates. Where the former posit that a theory must be constructed out of verifiable observational terms and theoretical terms, Popper regards the development of a theory as a non-rational affair that does not require any further logical justification (1959/1974, p. 31):

> The question how it happens that a new idea occurs to a man—whether it is a musical theme, a dramatic conflict or a scientific theory—may be of great interest to empirical psychology; but it is irrelevant to the logical analysis of scientific knowledge.

Thus, on the one hand, the creation of the product of knowledge, the theory itself, cannot be evaluated rationally. This logically implies that these products cannot be mutually compared with respect to their "truth". On the other hand, the falsification principle should encourage the growth of knowledge by means of the rational comparison of theories. In other words, the aim is precisely to produce rationally comparable theories after all. Popper escapes what seems to be a contradiction in the following way. He argues that the more a theory survives attempts to falsify it, the more it is regarded as being "corroborated" and therefore it is accepted, at least for the time being (Popper, 1963/1978, p. 231):

> Only with the help of bold conjectures we can hope to discover interesting and relevant truth.

In this, Popper's point of view (except for the terminology) approaches the logical positivists' idea of verification and confirmation. After all, the corroborated theory that has survived a number of tests in a deductive view of science is analogous to the confirmed theory that is "confirmed" by observations in an inductive view of science. This is the point at which logical positivism and critical rationalism meet, as will be shown more formally when discussing the theories on a methodological level. Popper himself also confesses that the reasoning regarding the growth of knowledge is problematic (1963/1978, p. 231):

> Looking at the progress of scientific knowledge, many people have been moved to say that even though we do not know how near or how far we are from the truth, we can, and often do, approach more and more closely to the truth. I myself have sometimes said such things in the past, but always with a twinge of bad conscience.

The problem concerning the growth of knowledge in relation to the falsification principle is solved to some extent in the next section in which confirmation and falsification are integrated. This relation is also important to psychology. Indeed, is falsification an effective practical strategy for the tester who wants to increase his knowledge? This is addressed in Chapter 3 onwards. Before I turn to this integration, a short review of the most important objections to verification and falsification in the philosophy of science is required.

Criticisms

The philosophies of science of both the logical positivists and Popper have produced libraries full of reactions. Feyerabend (1970), Hanson (1969), Kuhn (1973), Lakatos (1970), Laudan (1977), and Toulmin (1961) are some of the most prominent classical critics. In the present section, five critical arguments against the standard views are examined.

Most critics have a common point of departure. Put simply, this point of departure states that science *in reality* does not conform to the regulations or reconstructions of Popper or the positivists. The criticism is not aimed primarily at the inadequacy of the principles of verification and falsification as abstract methods of either establishing science rationally or enabling science to grow rationally. It simply denies that science actually occurs in accordance with rational guidelines. This group assigns a completely different role to the philosophy of science: The study of science should *describe* the dynamics of the scientific process, at the same time keeping in mind the psychological, sociological, and historical aspects, in other words, the "world view" of scientists (Suppe, 1977). The nature of scientific philosophical analyses and explanations of scientific developments is regarded by Kuhn (1970, p. 21) as:

> Already it should be clear that the explanation must, in the final analysis, be psychological or sociological. It must, that is, be a description of a value system, an ideology, together with an analysis of the institutions through which that system is transmitted and enforced. Knowing what scientists value, we may hope to understand what problems they will undertake and what choices they will make in particular circumstances of conflict.

These philosophies attempt to give a description of scientific conduct instead of a set of rules or norms (Bechtel, 1988). How does this approach regard the relationship between theory and evidence, and, moreover, how does it view the testing of theories?

One standard argument concerning the relationship between theory and evidence is that observations are not independent of theory. This argument affects verificationism and falsificationism alike. This viewpoint has been defended in depth by Hanson (1958) among others. Observation, according to Hanson, is determined entirely by the theory which one attaches to it. This reasoning can lead to the philosophical conclusion that rational theory selection with reference to observations is impossible. Although the argument that observations are imbued with theories has been defended in more or less adapted forms by most philosophers of science taking this approach, very few philosophers draw the radical conclusion that the testing of theories is totally impossible.

A less radical application of the argument is the view that the theory indicates which observations are *relevant* (Shapere, 1982). This version of the argument does not say anything about the influence of the theory upon that which one perceives, but deals with the influence of the theory upon the *selection* of that which the scientist must observe in order to test the theory.

The crucial distinction between this and the previous version of the argument is that, in the version in question, the theory has no influence upon the observation itself, that is, the result of the test. This second interpretation of the influence of theory on observation does not lead to relativism, in contrast to the first, since evaluation of the theory is possible in the light of independent observations. This version of the argument has some intuitive plausibility for the psychology of testing. Our beliefs obviously direct our attention to the facts we find relevant for evaluating our beliefs, but we are still able to observe discrepancies between our beliefs and inconsistent evidence (see Chapters 3 and 4). The relevance explanation of test selection has only recently attracted serious attention in the psychology of testing, although it is similar to what Shapere proposed for scientific reasoning.

The third critical argument against the views of the logical positivists and Popper is that theories do not stand alone; they are part of a more general theoretical system. Kuhn (1973) speaks of "paradigms", Lakatos (1970) of "research programs", and Laudan (1977) of "research traditions". It is characteristic of these theoretical systems that they cannot be falsified by just a single test. The choice between one theoretical system and another is not based on an empirical test of both systems. In one way or another, these systems are protected against refutation; they tend to resemble metaphysical assumptions. During the execution of everyday research within the system, the system itself is not brought into question (Lakatos, 1970, p. 177):

> Thus the "dogmatism" of "normal science" does not prevent growth as long as we combine it with the Popperian recognition that there is good, progressive normal science and that there is bad, degenerating normal science and as long as we retain the determination to eliminate, under certain objectively defined conditions, some research programs.

In a recent study on actual hypothesis testing by scientists, Gorman (1995) showed that these levels can be distinguished in scientific hypothesis testing. Thus, unlike the logical positivists and Popper, the critical philosophers of science feel that observations do not verify or falsify a theoretical statement "just like that". To them the broader theoretical context should be taken into account. It is because of that context that the "quest" for falsification is neither possible nor called for. Similar to the previous one, this argument may also be transposed to the everyday context: A hypothesis tester has some fundamental beliefs that are not readily rejected after one inconsistent observation (see Chapters 3 and 4).

The fourth critical argument against falsificationism comes from Kuhn (1970). He disputes the demand of falsificationism that for every theory it should be possible to establish which fact will refute it. This requires that the theory should be completely logically articulated, something that is never the case in practice. In the real world, theories state which facts will confirm them, or, in other words, they state what can be expected and what is irrelevant. That is what we want to be informed about, and that is why the theory has been framed. The falsification of the theory—if it happens at all—just happens: It cannot be described in advance. An interesting result of this point of view for psychology is that it is impossible to search for falsifications since they cannot be thought up in advance (see Chapter 3).

One last, much-stated argument against Popper's principle of falsification is that a theory is never refuted—nor should it be—in the absence of an alternative theory. This applies to individual theories, hypotheses, paradigms, and research programmes alike. Lakatos (1970) argues that the mere refutation of theories does not in any way benefit the growth of science, simply because of the fact that, after refutation, the researcher does not have a frame within which the necessary explaining can be done. This argument is basically psychological. From a cognitive viewpoint, it may be preferable to have a dubious theory about phenomena in our environment rather than to have no theory at all, even though this theory may be inadequate and possibly at odds with some phenomena.

To summarise, five arguments from the criticism of the standard theories of science, verificationism, and falsificationism, have been discussed. First, scientific knowledge acquisition is not governed by normative rules. Second, observations used to test a theory are determined or at least imbued by the theories they are supposed to test. Third, a theory is generally not rejected "just like that" as soon as a falsifying observation comes up. The general theoretical framework in which it is embedded protects it to some extent against rejection. The fourth problem with falsificationism is that it is not necessarily clear in advance which observations will refute a theory. Falsifications are facts that emerge rather than factors that are purposefully sought. Finally, some critics advance the opinion that theories are rarely rejected without an available alternative. Scientists as well as humans in general need explanatory theories to explain their environment.

In the next section, the focus is again on positivism and critical rationalism. I ask how incompatible they are. What remains of the discrepancies between these philosophies when we look at how they are concretised and formally elaborated?

METHODOLOGICAL PRINCIPLES

When a researcher tests a theory, the usual procedure is as follows: Hypotheses are derived from the theory and predictions are derived from the hypotheses. These predictions may or may not turn out to be true. Observation will show this. The relationships of the hypotheses to the theory and of the predictions to the hypotheses are similar; they are "logical deductions" or "empirical specifications" (de Groot, 1969). The deductive nature of the procedure outlined so far guarantees that it is logically correct: The reasoning moves from the general to the specific. However, when the prediction which is deduced from this reasoning turns out to be true, the theory is not proven correct, logically speaking. One cannot, when confronted with the correctness of a special prediction, reason backwards to the correctness of the theory. When the prediction does *not come true*, however, the theory is indeed logically falsified retro-actively. This reasoning corresponds to the logical version of falsificationism discussed earlier. However, scientists are often interested in the empirical tenability of their hypotheses and theories, rather than in single falsifications. Moreover, inductive reasoning is quite intuitive. For example, it seems quite obvious to regard a theory that has generated more predictions that have come true as being more plausible than a theory that has done this to a lesser extent. For this reason, a methodology that can justify the empirical correctness of hypotheses is a very useful instrument for scientists.

The history of philosophy has seen a number of attempts to approach the problem of induction and to construct a system of inductive logic. Carnap's *Logical Foundations of Probability* (1950), developed within the framework of logical positivism, is probably the most well known of these. Following Popper's impressive attack on induction, the attempts have been temporarily shelved. Salmon (1984, p. 27):

> It might have been felt that the problems in constructing an appropriate inductive or statistical model were so formidable that one simply did not want to undertake the task.

However, the revival of interest in inductive philosophy is striking (see Giere, 1975, 1977, 1988; Salmon, 1973, 1984). Moreover, Popper's formal elaboration of falsifying testing boils down to something that resembles an inductive theory. The essence of a number of theories of inductive confirmation of hypotheses will be outlined later, with particular emphasis on Carnap's theory (1950) in the section on confirmation. Popper's falsification principle in the testing of individual hypotheses will subsequently be discussed, and finally an integration is provided of these two main theories of induction and falsification.

Confirmation

Confirmation theories attempt to formalise the intuition that one can evaluate a theory or, more generally, a universal statement, on the basis of a limited number of observations, despite the fact that, speaking strictly logically, one cannot complete this kind of evaluation of a theory. Thus, a confirmation theory should indicate the extent to which a hypothesis is confirmed by certain observations. At the same time, a confirmation theory indicates the observations by which a theory should be tested in order to receive the highest possible grade of confirmation. The assigning of a value such as a degree of confirmation is called "probabilistic confirmation" (de Groot, 1969).

The most simple definition of the degree of confirmation (Erwin & Siegel, 1989) is the probability[1] of the hypothesis given the observation. Whenever this is large, in other words, as the value approaches 1, one can say that the observation x confirms the hypothesis H. This interpretation of confirmation can be found in roughly this form in the work of Carnap (1950) and Salmon (1975) (as the theory of absolute confirmation). This confirmation theory can be formally reproduced as follows:

$$\text{x confirms H if and only if } p(H \mid (x \text{ and } b)) > K \qquad (1.1)$$

where H is the hypothesis, x is the observation, b is the background evidence, and K is some criterion value. The criterion K is stricter the closer it comes to 1. Take, for example, the hypothesis: "The swans in the park are white". In order to confirm this hypothesis, we check 10 swans. The observations turn out to support the hypothesis: All 10 swans are white: $p(H \mid x$ and b) is very high. Suppose that we now check 100 more swans. The probability of the hypothesis that they are all white becomes even higher. However, the example illustrates a problematic aspect of this confirmation theory. If we already are pretty sure that swans are white animals, then all these confirmations do not greatly change this prior belief regarding H. But if we are completely uncertain about H, 10 observations can highly influence this belief. In other words, how much x confirms H also depends on H's prior probability.

An alternative interpretation of confirmation which tries to overcome this objection is the "relevance interpretation of confirmation". This was also formulated by Carnap (1950) and reintroduced by Salmon (1975, 1984). This posits that the degree of confirmation of a hypothesis increases the more the probability of the hypothesis, given the observation, exceeds the probability of the hypothesis without the observation. The observation, then, is more relevant for the hypothesis. Formally, this can be reproduced as follows:

$$\text{x confirms H if and only if } p(H \mid (x \text{ and } b)) > p(H \mid b) \qquad (1.2)$$

When we apply this confirmation theory to the previous example, it can be seen that the extent to which our new observations strengthen our belief

[1] Due to the fact that the probability refers to the hypothesis (and not to an observation), it concerns a "subjective" interpretation of probability.

depends on what we already believe on the basis of "old" background knowledge. Thus, this measure of the degree of confirmation takes into account prior knowledge on what is to be confirmed.

In order to establish the degree of relevance of the observation, Carnap formulated the relevance ratio:

$$\frac{p(H \mid x \text{ and } b)}{p(H \mid b)} \tag{1.3}$$

Thus, observation x is more relevant the more the ratio exceeds 1. The greater this is, the more an observation can be regarded as being a strong confirmation of the hypothesis. It will be shown in the next chapter that the most influential formal theory of confirmation, the Bayesian approach, uses a similar interpretation of degree of confirmation. In the Bayesian context, this ratio corresponds to the amount of "revision of belief". The common aspect of these definitions of degree of confirmation is a subjective view of probability, which can express and quantify degree of (subjective) belief (Howson & Urbach, 1989). How can this definition be translated into a guideline for the choice of test? According to the confirmation theories, we should perform the test that can produce a result that will confirm the hypothesis as much as possible. Thus, when selecting a test, a pre-posterior deliberation must take place concerning the expected degree of confirmation of the possible result, and the test that maximises this value should be chosen.

The relevance theory of confirmation comprises the idea that a theory should be subjected to a test that, in comparison to other tests, can produce observations that "belong more specifically" to the theory in question. The more this is the case, the better the theory can be confirmed. This indication can also be expressed as the requirement that the observation should be unexpected to the maximum degree possible in the light of mere background evidence. This general idea of unexpectedness and, at the same time, of being connected to the theory, is common to most theories of induction (Giere, 1977; see also Chapter 2). Notice that in the example of the swans, observing *all* swans in the park is the best empirical specification of the hypothesis "all swans in the park are white", because this observation is also most "specific" for the statement. If they all turn out to be white, the hypothesis is confirmed to the maximum extent.

In summary, in Carnap's work, the guidelines for testing are: Test the hypothesis against the prediction that is specifically associated with the hypothesis and that cannot be expected *without* assuming the hypothesis, because this one has the highest degree of confirmation.

We now turn to Popper's falsification standard, asking the same question as that posed in the confirmation theories: Which test should one use in order to evaluate one's hypothesis to the optimum extent?

Falsification

The logical advantage of the falsification principle is that a universal statement can be refuted by an opposite example. Accordingly, it is desirable to seek falsifications of the theory. Popper (1963/1978) postulates that the best way of achieving this is to test the theory as "severely" as possible. Popper (1963/1978) has designed a measure for the severity of a test. This is:[2]

$$S(x, H, b) = \frac{p(x \mid H \text{ and } b)}{p(x \mid b)} \tag{1.4}$$

Popper (1963/1978, p. 391):

> The severity of the test ["x" in the notation used here] interpreted as supporting evidence of the theory ["H" in the notation used here], given the background knowledge b.

According to this definition, a test is more severe the more the chance of the supporting observation occurring under the assumption of the hypothesis exceeds the chance of its occurring without the assumption of the hypothesis (i.e., under the assumption of background knowledge only). The more the ratio exceeds 1, the greater the severity of the test.

Let us look again at the example of the swans in the park. If we assume that the park contains only white swans (H), then the probability of observing 10 white swans is high: 1. However, the probability that 10 swans taken from the park will be white anyway is also quite high, given our background knowledge. The quotient is thus approximately equal to 1 and the test is not severe. But if we observe all the swans in the park, our expectation *a priori* that all will be white will decrease (the denominator), whereas the numerator is still 1. Actually, this is the most severe test we can think of. Thus, Popper wants scientists to choose severe tests to investigate their theories. If the theory "survives" such a test, a valuable proof of its empirical merits has been given.

Integrating confirmation and falsification

Surprisingly, however, the reason why a test is severe is basically the same as the reason why the test can provide a high degree of confirmation. Thus, if a test is good according to confirmation theory it is also good according to falsificationism, as I will show later. Indeed, it can be demonstrated that the

---

[2] Note that, in contrast to the confirmation theories, the probability p refers to the observation and not to the hypothesis. This means that the probability is not subjectivistic but frequentistic.

measure of the severity of the test and the relevance ratio (Carnap's measure for the degree of confirmation) are equivalent.

Proof 1.1

To reduce the severity of test definition to the relevance ratio, we use Bayes' theorem (see also Howson & Urbach, 1989). This theorem is the following equation:

$$p(H \mid x) = \frac{p(x \mid H) \cdot p(H)}{p(x)}$$

Now, we rewrite the definitions (1.3) and (1.4) without the background term b. We can do this because b is whatever is known and not put into question. It is the background knowledge in (1.4) and the background observations in (1.3). b is assumed to be true in both (1.3) and (1.4). Thus, its probability is 1. Accordingly, we rewrite the nominators and denominators of (1.3) and (1.4) as follows:

$$p(H \mid b) = p(H)$$
$$p(H \mid x \text{ and } b) = p(H \mid x)$$
$$p(x \mid b) = p(x)$$
$$p(x \mid H \text{ and } b) = p(x \mid H)$$

The severity of an observation S can be rewritten by means of Bayes' theorem:

$$\frac{p(x \mid H)}{p(x)} = \frac{p(H \mid x)}{p(H)}$$

which is the relevance ratio.

The parallel between the relevancy interpretation of confirmation and the severity of the test is even more striking when Popper derives "degree of *corroboration*" from the degree of severity of the test. The function C indicates the degree of corroboration of a theory caused by datum x.

$$C(H, x, b) = S(x, H, b) \tag{1.5}$$

This degree of corroboration (function C) is equivalent to the relevance ratio. This can be seen when one orders the terms (between parentheses) of S as prescribed in C. Notice that the probability p in the function C, by virtue of the order of sequence of the parameters that Popper assigns to C, refers to the hypothesis. Thus the probability in C must be interpreted as "amount of belief" rather than frequentistically, precisely as in the relevance ratio. By introducing the degree of corroboration, Popper brings the severity principle back to Carnap's relevance ratio. Thus, a given observation has the same value for a given hypothesis according to the falsificationist standard (i.e., severity of test) as it does according to the verificationist standard (high degree of confirmation) although both standards issue from conflicting philosophies.

De Groot (1969) gives a number of indications for test choice, which aim at maximising the relevance as well as the severity of a test. I shall cite them briefly here because they provide a good informal idea of the commonality between the principles of confirmation and severity of test. De Groot (1969) points to the importance of selecting a "relevant prediction" from all the predictions generated by the theory. This is the one that should be tested. He distinguishes four factors that determine the relevance of a prediction for evaluating a given theory.

First, the relevance of a prediction is determined by the degree of particularisation that occurs between the theory and the prediction. The prediction may refer to only a very small component of the theory. In that case, there is only a small part of the theory at issue in the test. Accordingly, the test is neither very relevant nor severe for the whole theory, and the degree of confirmation does not increase substantially if the test is passed. The second factor is the degree of precision of the prediction: A prediction can be precise or "vague". Third, the prediction is more relevant the more it has been deduced from a fundamental premise in the theory. The more this is the case, the more this helps confirm the theory, but it also means that the test is more severe. The final important factor mentioned by de Groot is the extent to which the prediction is critical; in other words, the extent to which it can separate a theory and an alternative theory. In general, a prediction is critical when it is able to refute other plausible theories or hypotheses. According to de Groot (1969, p. 107),

> The areas of conflict between two rival theories often afford good starting-points for the articulation and empirical (experimental) realisation of relevant predictions.

In the next section, I shall evaluate the contribution of the previous analysis to the psychology of testing behaviour. In addition, some comments will be made on the falsification standard in practical test situations, and I shall also return to the distinction between logical and psychological falsification.

IMPLICATIONS FOR TESTING BEHAVIOUR

What are the implications of the joint principle of severity and relevance for testing behaviour? First, a few examples of what severe testing would look like in a concrete testing situation are given. Second, a paradox in Popper's falsification standard is pointed out, which arises from the combination of test severity and the high value he assigns to refutation. In the final section, I return to the psychological version of falsificationism.

Confirmatory behaviour is falsifying behaviour

It has been shown that the formal definitions of relevance and severity are equivalent in spite of the highly contrasting philosophies that have produced them. This implies that a given prediction of a given hypothesis is just as "good" a test according to the confirmatory standard as it is according to its rival, the falsificationist standard.

How can one choose good tests in practice? De Groot's determinants of relevant predictions form a useful guideline. Let us consider two examples. Suppose, for example, that a physician wants to test whether or not his patient has disease X. Two kinds of bacteria, A and B, are equally prevalent, but bacterium A is seen more often in combination with X than is bacterium B. In such a case, testing for A is a more severe test of the presence of illness X than testing for B. Testing for A represents a stronger "particularisation" from theory to prediction, because A occurs more specifically in cases of illness X. Also, B is less "fundamental" to X than A. Formally: The denominator in the severity of the test is the probability of the observation. This quantity is equal for both predictions since they have equal prevalence. But A has a higher conditional probability under X (the numerator). Therefore, the severity of test for A is higher than the severity of test for B. Suppose a hypothesis deals with someone's personality, e.g., "John is introvert". A severe test may consist of asking John whether he prefers to spend all his weekends alone. A less severe test would consist of asking whether he prefers reading a book to attending a concert. Indeed, both observations are likely, if John is introvert. But, in de Groot's terms, one could argue that spending all one's free weekends alone is a more fundamental characteristic of introversion than preferring reading to concerts.

An important implication of the unified measure for severity of test and degree of confirmation, is that it is inversely related to the prior probability of the evidence (and of the hypothesis in subjectivist theories of confirmation). This inverse relation will play an important role in the next chapters because it is intuitive, and for this reason very important for psychological models of testing behaviour. Howson and Urbach (1989, p. 86) also point out this intuition.

> These facts are reflected in everyday experience that information that is particularly unexpected or surprising unless some hypothesis is assumed to be true, supports that hypothesis with particular force.

In other words, both the severity and the relevance of a test increase as the observation is more strongly related to the hypothesis on the one hand, and *a priori* more unexpected on the other. A tester behaves as a rational confirmer *and* falsifier if these kinds of observations are pursued.

A Popperian paradox

The analysis of Popper's falsification standard made in the previous section has a singular consequence for the hypothesis tester. Indeed, according to Popper, the falsificationist attitude basically contains two rules of behaviour. On the one hand, the hypothesis tester should wish the tests to be as severe as possible, in order to maximise the degree of corroboration of the theory. Popper (1963/1978, pp. 240–242):

> A serious empirical test always consists in the attempt to find a refutation . . . We require that the theory should pass some new, and severe, tests.

On the other hand, the tester should be satisfied if the test indeed results in a refutation. Popper (1963/1978, p. 243):

> Refutations have often been regarded as establishing the failure of a scientist, or at least of his theory. It should be stressed that this is an inductivist error. Every refutation should be regarded as a great success; not merely a success of the scientist who refuted the theory, but also of the scientist who created the refuted theory . . . .

However, within the Popperian framework, these two attitudes are contradictory, as I shall show. Indeed, the severity of a piece of evidence is equivalent to its corroboration value. Thus, if a test is more severe, it will corroborate the hypothesis better, if indeed the hypothesis stands the test. But this severe test can also result in a falsification, of course. It is even stronger. The more severe the test, the more *likely* such a falsification is. Also, as the possible confirming outcome of a given test can corroborate better, the possible falsifying result falsifies worse. That is, it has little power to prove the hypothesis to be false. Hence, the *value* of a *refutation* of a test is inversely related to its severity, and hence to the corroborating *value* of its *supporting* outcome. This can be shown formally for a simple case.

Suppose two tests X and Y for H, each with two possible outcomes: x and y (supporting H) and x' and y' (refuting H). Test X is more severe than test Y with regard to hypothesis H. Thus:

Proof 1.2

$$\frac{p(x \mid H \text{ and } b)}{p(x \mid b)} > \frac{p(y \mid H \text{ and } b)}{p(y \mid b)}$$

This implies:

$$\frac{1 - p(x \mid H \text{ and } b)}{1 - p(x \mid H)} < \frac{1 - p(y \mid H \text{ and } b)}{1 - p(y \mid H)}$$

Thus,

$$\frac{p(x' \mid H \text{ and } b)}{p(x' \mid b)} < \frac{p(y' \mid H \text{ and } b)}{p(y' \mid b)}$$

So, the more severe the test, i.e., the better the supporting result can corroborate the hypothesis, the less its potential to falsify the hypothesis. In other words, if we choose severe tests in line with the Popperian prescription, we shall obtain either strong supporting evidence or weak refuting evidence. The refuting evidence is less powerful as the test is more severe. A severe test only leads to satisfaction when supporting evidence is observed. Therefore, there is little reason to be satisfied if we obtain a refutation from a severe test.

This can be illustrated again informally with the example of John's introvert personality. Indeed, if John turns out *not* to spend all his holidays and weekends alone, we cannot be certain that he is not introverted. He can still be pretty introverted without being alone every weekend! We just cannot be sure. The test has not improved our knowledge very much. Thus hunting falsifications (testing severely) is nice as long as one does not find them, because supporting evidence from a test which was directed at finding refuting evidence is more informative about the hypothesis than the actual refuting outcome. It is thus the combination of *seeking* refutations *a priori* and considering them as a "great success" *afterwards* that is contradictory in Popper's normative theory.

Falsification as moral standard

We can regard falsification and confirmation either as logical theories or as behavioural standards. That is, Carnap and Popper not only proposed these principles to assess the quality of a test, but they also prescribe that the hypothesis tester should maximise them, that is, try to falsify or obtain better confirmation. In Popper's theory, falsificationism has a kind of moral force, as was argued above (1959/1974, p. 50):

> A system such as classical mechanics may be "scientific" to any degree you like; but those who uphold it dogmatically—believing, perhaps that it is their business to defend such a successful system against criticism as long as it is not conclusively disproved—are adopting the very reverse of that critical attitude which in my view is the proper one for the scientist.

The "good" tester in psychological falsificationism tests his or her *own belief* as severely as possible instead of "the theory". This means that the methodological instruction always to test as severely as possible no longer applies when the researcher does not believe in the theory to be tested. Particularly in the case of two rival theories, the researcher should look for confirmation of the theory which is untrue according to his or her opinion. Only then does the behaviour comply with the rules of psychological falsification. Also, the position of the tester determines the expectation concerning the result of the test. De Groot (1969, p. 101):

> Anyone wishing to refute a given general hypothesis [like "All A's are P"] will undoubtedly look for cases which are non-P, but, anyone wishing to prove it will do the very same thing, albeit in hopes of not finding them!

The distinction between the logical and psychological approaches of falsificationism relates to the distinction between cognitive versus motivational biases in hypothesis testing in the psychology of testing behaviour. On the one hand, psychologists examine whether testers know the logical principle of falsificationism and use it spontaneously. On the other hand, one can study the extent to which testers attempt to refute their own beliefs. The interesting question then concerns whether or not testers apply conservative tests in order to protect their own views from refutation. This is the motivational aspect that is mostly studied in social hypothesis testing (Trope & Liberman, 1996). In early empirical research, the logical and the psychological aspects involved in testing in a falsifying manner were often equated. That is, testing in a logically erroneous way was often interpreted as being caused by conservatism with regard to the hypothesis. This linkage between the logical and the "moral" inferiority of confirmation might have been directly adopted from Popper's critical ideas. Wason (1960, p. 139):

> But the readiness (as opposed to the capacity) to think and argue rationally in an unsystemized area of knowledge is presumably related to other factors besides intelligence, in so far as it implies a disposition to refute, rather than vindicate assertions, and to tolerate the disenchantment of negative instances. And certainly these qualities are no less important for thinking in general than the more obvious cognitive functions associated with purely deductive reasoning.

However, the claims authors make about hypothesis testing have become less "moralistic" with the passage of time. In addition, there is an intuitive problem with motivationally biased testing behaviour (Poletiek, 1996). Indeed, the intention to subject an idea to a test is, to some extent, incompatible with the intention to prevent it from being "really" tested, that is, to gather new information about the idea. We will come back to this problem later. Here we want to stress that motivational explanations of testing biases are hard to account for, merely on intuitive grounds.

SUMMARY AND CONCLUSIONS

In this chapter, a selective review has been given of a number of theories from the philosophy of science concerning testing. The central question which was asked was: What can be accepted as being a good test? Accordingly, these theories have been examined concerning the indications they suggest for the tester who wishes to subject ideas to scrutiny. The two most important classical competing theories in the philosophy of science—verificationism and falsificationism—appear to be equivalent both formally and in terms of behaviour. A good test is one that may lead to a relatively large degree of confirmation of the theory (according to verificationism). But in doing so, it necessarily forms a severe test (which is the quest of fasificationism) as well. Nonetheless, it is falsificationism and not verificationism that has served as a theoretical and normative basis for the psychology of testing behaviour, probably due to the open-minded and anti-dogmatic character Popper gave to his own philosophy.

The major findings of this chapter are that confirmatory testing and falsifying testing, as defined in the great philosophies of science, boil down to a similar model for testing behaviour. Also, it has been shown that the value of test outcomes for evaluating hypotheses can be assessed, and that it relates to their prior probability. In Chapters 3, 4, and 5, the value of this "unified" model for hypothesis-testing behaviour will be addressed. However, before this, I shall turn briefly to another set of theories that have often been used as a basis for the study of testing: the formal theories of testing. The questions again will be: How do the formal theories of testing represent the testing process? What is considered a good test according to these theories; and how do they relate to our present analysis of the philosophical theories?

CHAPTER TWO

Formal theories of testing

INTRODUCTION

In formal theories of testing, procedures of testing are systematically described, and the test procedure as well as the results are quantified. These theories are either descriptive or normative, or encompass both elements, just as in the philosophies of testing. Furthermore, they are frequently used as models for testing behaviour (Klayman & Ha, 1987; Oaksford & Chater, 1994b; Poletiek & Berndsen, 2000; Skov & Sherman, 1986; Slowiaczek, Klayman, Sherman, & Skov, 1992). The most commonly used mathematical model for describing testing behaviour is Bayes' theorem (see later section: Fischhoff & Beyth-Marom, 1983; Kirby, 1994; Klauer, 1999; Oaksford & Chater, 1994b; Skov & Sherman, 1986; Slowiazcek et al., 1992). A few researchers have based their studies on information-theoretic models (Oaksford & Chater, 1994b; see later section). The "decision making" theories of testing (next section), such as significance testing and Neyman and Pearson's theory, are rarely used (Friedrich, 1993) in spite of the fact that they form the standard testing procedure in most social and behavioural sciences.

The goal of this chapter is to provide an overview of these theories, and to evaluate their relevance for everyday hypothesis testing. The focus will be on the conceptual similarities between the theories. They will also be compared to the philosophical theories expounded in the previous chapter. Departing from this comparative analysis, I shall evaluate their ability to contribute to the understanding of human testing behaviour. I start with the decision making models.

TESTING AS DECISION MAKING

Deciding the truth of hypotheses

The theories discussed in this section assume that hypothesis testing is a process of *deciding* about the truth of hypotheses. The typical situation is that a hypothesis tester wants to know whether or not a given hypothesis H is true. A decision is made, on the basis of some information, to "accept" or "reject" the hypothesis. However, this information is incomplete. It cannot *definitely* decide the truth status of the hypothesis, as is the case in proposition logic (see Chapter 4). Indeed, if it could, no "decision" would have to be made; instead, evidence would simply definitely "reveal" whether or not the hypothesis were true. The tester needs a procedure for making this decision. In the present framework, testing is basically looked upon as choosing this decision rule. The situation can be represented in the famous trade-off matrix (Coombs, Dawes, & Tversky, 1970) displayed in Table 2.1(a). The tester can make either a right or a wrong decision, which corresponds to the situations on the two diagonals of the matrix.

Originally, decision making theory was applied to model *actions* depending on perceived states. For example, in signal-detection theory, this model describes how a human responds to ambiguous stimuli which correlate with the state of the world. In scientific hypothesis testing, the stimuli correspond to the incomplete "sample" information, and the responses indicate the acceptance or rejection of the hypothesis about the state of the world. Interestingly, considering hypothesis testing as being decision making implies that evaluating the truth of hypotheses has something arbitrary or subjective about it. Deciding about the truth of a statement has even been described as a contradiction (Hays, 1973). According to Friedrich (1993, p. 298):

> A statement is true or false in reality independently of human decisions. However, the assumptions that hypothesis testing is generally based on incomplete knowledge and depends to some extent on subjective decisions of the tester, are quite adequate for pragmatic testing situations.

Indeed, in most everyday test situations, the tester will be working with incomplete information and will have subjective feelings about the damage incurred in the event of a wrong decision having been made regarding the hypothesis.

TABLE 2.1
Trade-off matrix for deciding about the truth of one hypothesis H (a), and for deciding between two hypotheses (b), with the probabilities of each situation

	State of nature	
(a) Decision		
	H is true	H is not true
Accept H (do not reject H)	$1-\alpha$	β
Reject H	α	$1-\beta$
(b) Decision		
	H_1 is true	H_2 is true
Reject H_2 (accept H_1)	$1-\alpha$	β
Reject H_1 (accept H_2)	α	$1-\beta$

Two theories within this framework are presented: Fisher's well-known significance test, and the theory of Neyman and Pearson. As the theories, discussed later, are formal theories, the hypothesis is often reduced to a "statistical hypothesis" and the incomplete information is assumed to be sample information or some other indirect information. These restrictions, however, are not a fundamental impediment to their evaluation as possible models for human behaviour.

Significance testing

In behavioural science, a frequently used method of testing hypotheses is the test of significance, developed by Fisher at the beginning of this century (see also Hays, 1973; Krueger, Daston & Heidelberger, 1987; Krueger, Gigerenzer, & Morgan, 1987). Its logic runs as follows. The (statistical) hypothesis to be tested is formulated. Then a test statistic is selected which is a relevant valuator of that hypothetical value and of which the sample distribution can be calculated. This is the probability distribution of the possible observations, presuming that the hypothesis is true. A "rejection area" is chosen within the sample distribution. This means that if the testing quantity assumes a value within that area, it will be decided that the hypothesis is *rejected*. This area should be "far from" the values that are expected when the hypothesis is true and the parameter value is right. Observations within the rejection area should be very improbable under the "reign" of the hypothesis.

Fisher states that, given the hypothesis, the probability that this particular observation will be made, leading to rejection, may not be greater than $\alpha = 0.05$. It refers to the situation in which the hypothesis is "rejected" but is nonetheless true (Table 2.1a). Since he proposed that value, it has become the conventionally accepted rejection area by most scientists using significance tests.

Finally, the test is actually performed and the observations made. The result may fall either inside or outside the rejection area, and a decision about the truth status of the hypothesis is taken accordingly. That is, the hypothesis is rejected or not rejected (Hays, 1973). Thus, the test of significance is explicitly oriented towards rejection. Fisher (1959, p. 42) states:

> A test of significance contains no criterion for "accepting" a hypothesis.

Interestingly, Fisher's theory of testing seems to resemble Popper's falsificationism. I shall go into this comparison briefly. Giere (1975) stresses, as follows, the similarity between Fisher's theory of testing and Popper's theory. Fisher, in Giere (1975, p. 225):

> Every experiment may be said to give the facts a chance of disproving the null hypothesis.

Giere even states that the large influence held by Fisher's theory at that time accounts for the favourable reception of Popper's ideas in the English scientific community.

This quote also indicates, however, the point at which the comparison between the falsification principle and the significance test breaks down: Its goal is to disprove the "*null*" hypothesis. The hypothesis to be tested in a significance test has a rather special character. The null hypothesis is usually the complement of the hypothesis that the researcher wants to evaluate. This interpretation of the hypothesis to be tested in the Fisher theory is not dictated by procedure but is commonly used in science. One reason for starting with the null hypothesis is that, in scientific work, it is assumed, *ceteris paribus*, that the researcher is wrong and the null hypothesis is correct. The null hypothesis is only rejected if there are cogent, empirical reasons to do so. The cogency of the empirical information depends on the decision on the critical area, which Fisher proposes making very restrictive. Thus, on the one hand, the restrictive rejection area may be seen as a Popperian aspect of the significance test, because the researcher has an arduous task in rejecting the null hypothesis. On the other hand, the requirement of trying to reject the null hypothesis, being the one the researcher does not favour, is un-Popperian since it actually requires a confirmatory attitude with regard to the theory believed to be true.

Is the significance test a suitable model of human hypothesis-testing behaviour? Before looking at the empirical research, we can think of a few arguments supporting the idea that humans proceed along the lines of the significance test when testing ideas in everyday life (Farris & Revlin, 1989), but there are also arguments against this. For example, the assumptions that (1) hypothesis testing is basically making a decision and that (2) the evidence used is incomplete are quite common in everyday reasoning. However, the focus on rejection is not very plausible in real-world reasoning, and neither is the severe restriction area. That is, there is obviously not anything like a conventional rejection area in everyday reasoning. The rejection area probably varies, depending on the pragmatics of the situation. Also, most everyday testing problems aim at accepting a hypothesis. Generally, we want to know whether or not, or to what extent, something is "true". The Neyman and Pearson theory of testing partly meets this problem. It assumes that the tester has *two* hypotheses, between which a choice must be made.

Deciding between two theories

The Neyman–Pearson theory of testing differs from Fisher's approach on two main points. First, not one but two hypotheses are put to the test. The underlying philosophy is analogous to that of Lakatos, who claimed that in a scientific test situation there are always two theories that contend for the facts. The testing situation is represented by Neyman–Pearson as a situation in which a choice must be made between two hypotheses on the basis of some information. The second point is that the criterion used for the choice between the two hypotheses is determined in such a way that the tester can control, to some extent, the risk of making incorrect decisions.

The Neyman–Pearson testing procedure is as follows: Two statistical hypotheses are formulated which are quantified predictions of the theories that are to be tested (H_1 and H_2). H_1 is the hypothesis to which α applies (analogous to the Fisher type of procedure), and the alternative H_2 is the hypothesis of which the β (the chance of wrongly rejecting the alternative hypothesis) can be calculated; see Table 2.1(b). Thus, a critical area is designated (the value of α is set) so that when the observed experimental results show a value for the testing quantity in this area, the decision to reject the first hypothesis is made; and when this value is not within the critical area, the decision is made to reject the alternative hypothesis.

Of crucial importance in the Neyman–Pearson theory is the specification of the rejection area. The chosen area determines the probabilities of the two decision errors a tester can make. Minimising one of them implies increasing the other type of error. Specifying the critical area represents determining the risk one is willing to take of making those errors. There is a direct relation between the "utility" (subjective costs and benefits) one attributes to these errors and the α and β probabilities that one chooses. It is demonstrable that a test can be described either in terms of these "utilities" or in terms of the "probabilities" attributed to the two errors. This can be felt intuitively as well. If one type of error implies high costs, the tester will avoid them by all means available: For example, the erroneous rejection of H_1 might imply high costs; the criterion value of α will be kept very small. Thus, the low probability of α means a highly negative utility of the corresponding decision error. When in doubt, the tester will mostly decide to accept H_1. The "price" for this strategy is, however, that H_1 will often be accepted erroneously (β increases). Thus, the Neyman–Pearson test procedure tolerates a clear subjective influence of the tester on what will be accepted as the truth. The tester has a certain amount of control over the posterior decision on what is true or false. Of course, there is the influence of the facts as well, and the less ambiguous they are, the fewer wrong decisions there will be. This corresponds to the situation in which the observations correlate highly with the true state of affairs, or situations in which a great number of observations can be made. But given that the quality of the observations is fixed, it is left to the tester as to how the decision is made concerning the acceptance of both hypotheses.

A crucial feature of the Neyman–Pearson theory is the method based on the Neyman–Pearson lemma—to choose the "best test", that is, the best decision rule, given that one has made up one's mind about the utility of making a decision error with regard to H_1 (i.e., rejecting it erroneously). Here, only the idea of the lemma is pointed out, not its mathematical proofs. In brief, for a certain set of observations and two hypotheses, it calculates the smallest possible value of β given a certain chosen value of α. Thus it gives the tester the best test, defined as the decision rule that will minimise the number of erroneous acceptances of H_1, given the number of false rejections of H_1 we agree to tolerate. In other words, it limits, as much as possible, the "price" of false acceptances of H_1, which we have to pay for the degree of reliability we demand with regard to decisions about H_1. This is achieved by calculating the "likelihood ratio" of the test. This ratio expresses the probability of a test result occurring, given that the hypothesis is true (in the numerator), and that the alternative is true (the denominator): $p(x \mid H_1)/p(x \mid H_2)$. In the Neyman–Pearson framework, this ratio can be interpreted as the ratio of the probability of making a correct decision with regard to H_1 ($p(x \mid H_1)$) divided by the probability of making an incorrect decision with regard to H_2 ($p(x \mid H_2)$). The likelihood ratio is a key term in most mathematical theories of testing. It will be discussed in more detail in relation to other theories of testing. The use of this ratio in the lemma is explained as follows (Hoel, 1984, p. 230):

> The conclusion of the theorem seems intuitively very reasonable because it essentially tells one to place in the critical region the sample points that most highly favour H_2 over H_1, that is, where the probability density under H_2 is very large compared with that under H_1.

In a way, the Neyman–Pearson lemma is normative. It tells us what is the "best test", that is, the best decision rule, given that our tolerance with regard to one type of error has been fixed.

One last implication of the Neyman–Pearson model discussed here is the asymmetry between a confirming and a falsifying outcome of a given test: If we reduce the rejection area, an observation in this area is very unlikely under the condition of H_1 being true. Thus, we can be confident of rejecting the hypothesis *if* such a result is obtained. However, if we do not observe such a result, we cannot reject H_1, but we will intuitively be less sure that H_1 is actually true and H_2 not. Lindley shows that the α-probability is often interpreted as expressing the probability that H_1 is true or false if sample observations fall within or outside the rejection area. This is, mathematically speaking, an incorrect interpretation, since probability in the decision making framework is frequentistic (see Lindley, 1965). However, although the decision making model does not allow the formal expression of the extent to which a hypothesis should be believed after it is tested, we almost automatically make this type of psychological inference about the confidence we should have in the accepted hypothesis. There is, thus, some asymmetry between the value of confirmations and falsifications, issuing from the same test, in the Neyman–Pearson model. This is consistent with our analysis of severe tests and the relevance ratio in test philosophies.

The assumptions underlying the Neyman–Pearson theory are very intuitive for everyday situations. Indeed, the Neyman–Pearson theory represents testing an idea in the same way as making a decision about our environment with the objective of acting upon this environment. For that reason, the Neyman–Pearson approach might be seen as an evolutionary approach to hypothesis testing. On the one hand, the utilities represent the fact that the tester is not neutral with regard to the evidence, but has an interest or some goal to pursue. To stay within the evolutionary metaphor: The utilities may be the survival principle. They determine which errors we can and cannot afford. On the other hand, the evidence represents the "hard" environment. Neyman–Pearson regards the practical goals of a tester as part of the hypothesis-testing process. In comparison with the Fisherian model, this has some advantages. In contrast to Fisher's procedure, the Neyman–Pearson procedure leads to *something* eventually being accepted. Another difference is that the Neyman–Pearson model gives a psychological interpretation to the α-criterion. Rather than representing a quite arbitrary rejection area, Neyman–Pearson defines it as the willingness of the tester to tolerate a certain amount of wrong decisions. The trade-off matrix (Table 2.1) and the dilemma it implies for a tester are very close to our intuitions. The more one wants to avoid one type of error, the more one will have to tolerate errors of the other type. But also the influence of the objective factor in the testing situation is easily understandable: The less ambiguous the information is (thus the more reliable one demands the data to be), the less one will make errors about H_1 and H_2.

To round off the discussion of the Neyman–Pearson procedure, I examine, using two examples, the conditions under which an everyday testing situation will be more or less suited to a Neyman–Pearson analysis. Consider the case where a tester has two hypotheses about his or her defective television: The cause is either a small defect that the tester can repair him- or herself or it is a serious defect that needs to be repaired by an expert. Deciding erroneously that the television has no serious defect when it does, implies that the tester will waste lots of time trying to discover the defect in order to repair it. Deciding erroneously that the defect is serious when it is not, will lead to incurring high costs for something the tester could easily have done him- or herself. Setting the decision rule will depend on the utilities our tester attributes to time and money, and the decision about which hypothesis is true will depend partly on this subjective trade-off. The evidence-gathering consists of scrutinising the interior of the television, or taking it to an expert. This person's testing behaviour could be adequately predicted on the basis of the Neyman–Pearson model, when his utilities are known. However, suppose that an archaeologist wants to know whether or not an excavated object is from the pre-Christian era. The archaeologist examines the object and decides on its date. The utilities are problematic in this example. The tester may have few feelings about the relative losses entailed in making each kind of error. An analysis in terms of Neyman–Pearson is more difficult.

The Neyman–Pearson theory of testing is quite adequate for describing testing behaviour when it resembles a pragmatic setting in which the tester checks the state of nature to find out what *to do*. However, the more the testing situation aims at solely discovering the "truth", the less it fits the Neyman–Pearson model. This happens when utilities cannot clearly be attributed to the errors, as, for example, in the more scientific situation mentioned previously. Nonetheless, decision theory is common in scientific discourse. But in that case, the utility problem is solved by introducing the convention of attributing a very high loss value to falsely rejecting the "null-hypothesis". This conventional solution still reflects the friction that results from the application of a decision making model to "pure" empirical hypothesis testing.

Bayesian theory is an alternative to the decision making approach. It considers hypothesis testing as an inference problem rather than a decision problem. According to Lindley, it meets the shortcomings of the decision making models, especially when applied to science-like reasoning situations (Lindley, 1965, p. 67):

> The person making the inference need not have any particular decision problem in mind. The scientist in his laboratory does not consider the

decisions that may subsequently have to be made concerning his discoveries. His task is to describe accurately what is known about the parameters in question.

THE BAYESIAN APPROACH TO TESTING

Testing as revision of belief

Bayes' theorem calculates how to infer knowledge about a hypothesis both from prior knowledge about it and from some new evidence. Phillips (1973, p. 5) describes Bayesian inference as follows:

> Opinions are expressed in probabilities, data are collected, and these data change the prior probabilities, through the operation of Bayes' theorem, to yield posterior probabilities.

Interestingly, the Bayesian approach can be seen as a formal confirmation theory, described in the previous chapter. In the Bayesian framework, as well as in philosophical confirmation theory, evidence confirms or updates a belief to the extent p(H | evidence) exceeds p(H) (Howson & Urbach, 1989). Thus testing is represented as hypothesis evaluation, which involves not only evidence regarding the hypothesis but also prior beliefs. The formal expression of the theorem is:

$$p(H \mid x) = \frac{p(x \mid H) \cdot p(H)}{p(x)} \tag{2.1}$$

In the theorem, Phillips' formulation of the Bayesian procedure can be recognised. The right-hand side of the equation expresses the prior probability attributed to the hypothesis (p(H)) and the influence of the data is expressed in $p(x \mid H)$ and p(x). Those elements are combined and provide the posterior probability of the hypothesis after the evidence is gathered: $p(H \mid x)$. Notice that the probability function p applies to the hypothesis as well as to the evidence in the Bayesian framework. In the former case, the probability means the amount of subjective belief that is attributed to the hypothesis. In the latter case, the probability function is frequentistic: It corresponds to a proportion. The subjectivist interpretation of probability and the introduction of quantified prior beliefs in the inference process form the core difference between Bayesian reasoning about testing and decision making approaches. This is also the basis of scholastic debates between the two frameworks, but these are not addressed in depth here.

Clearly, the subjective element introduced in the testing process refers to the prior probability of the hypotheses. Analogously, the decision making framework allows the use of subjective utilities of decision errors. A hypothesis tester in the Bayesian model is represented as a person wanting to update, by means of some information, his or her subjective beliefs with regard to a hypothesis. The relative influence of evidence and belief depends on the quality of the evidence on the one side, and on the degree of prior belief on the other. As the data gathered in the test are more numerous or less ambiguous, their influence will increase. Conversely, strong prior beliefs will increase the relative influence of those beliefs. The theorem (2.1) can also be written out in terms of two hypotheses being tested against each other. This version is called the "odds likelihood ratio" version.

$$\frac{p(H_1 \mid x)}{p(H_2 \mid x)} = \frac{p(x \mid H_1)}{p(x \mid H_2)} \cdot \frac{p(H_1)}{p(H_2)} \qquad (2.2)$$

The likelihood ratio is the first term on the right-hand side of equation (2.2). It describes the properties of the evidence for assessing the hypotheses H_1 and H_2 in relation to each other. In the Bayesian framework, a test brings about the most revision of belief if the ratio $p(x \mid H)/p(x)$ is maximised, or the likelihood ratio $p(x \mid H_1)/p(x \mid H_2)$ maximally deviates from the value 1. Phillips (1973, p. 83) states:

> We can say then that the size of the likelihood ratio determines the amount of revision of opinion. Data are informative if they lead to a very large or very small likelihood ratio, they are non-informative if the likelihood ratio is near or equal to 1.

Relation to other theories of testing

The Bayesian approach to testing has interesting parallels with other approaches discussed previously. Apart from the technical aspects and the underlying fundamental differences, e.g., with regard to the concept of probability, the Bayesian approach shares many implications for testing behaviour with both the other formal and the philosophical perspectives. First, the term describing the properties of the evidence (equation 2.1) used to test the hypothesis is equal to the definition of Popper's test severity. In the previous chapter, it was shown that maximising the severity of a test leads to a high degree of corroboration if the test succeeds. An analogous relation can be demonstrated in the Bayesian framework. Maximising the ratio in (2.1), while keeping the prior beliefs constant, implies maximisation of belief revision.

Second, there is a relation between the impact of the evidence and the prior beliefs in the Bayesian approach on the one hand, and the setting of the decision criterion in the Neyman–Pearson approach on the other, even though both theories are based on totally different principles. Bayes assumes that the prior subjective probabilities are independent of the likelihood ratio of the evidence. Now, the likelihood ratio can be transcribed in decision making terminology as follows. Suppose the observation x to be the criterion value for rejecting H_1. The probability of observing a value in the rejection area delimited by x, given that H_1 is true, is α. The probability of observing a value in this area when H_2 is true is $1-\beta$ (cf. Table 2.1). Thus the likelihood ratio for such an observation is $\alpha/(1-\beta)$. The decision criterion x in the Neyman–Pearson context was obtained by setting the values of α and β. This was done on the basis of utilities of decision errors, and, if one wishes, with the help of the Neyman–Pearson lemma. Thus, in the decision making framework, the choice of this criterion is not necessarily independent of prior beliefs in the two hypotheses (Gigerenzer & Murray, 1987). This choice may involve all kinds of considerations, including considerations about prior belief in the hypotheses, and the utilities attributed to the two types of error.

For example, a sailor wants to test the hypothesis that the wind will be suitable tomorrow, with the idea of putting to sea. The yacht is unreliable in stormy weather. Since it has been very windy up until today (force 7 on the Beaufort scale), the tester believes *a priori* that there is a very high chance that the wind will also be force 7 or higher tomorrow. Therefore, he or she decides to leave the port tomorrow only if the weather forecast predicts a wind of force 5 or less. The rejection area is chosen in order to minimise the probability of a wrong decision with regard to a storm. However, the sailor's prior subjective belief about the wind force might well have been different if, for instance, it had been very calm for the last week. In line with this prior belief, the sailor might well have shifted his or her criterion to force 6. This shift of the decision criterion implies a shift in the values of α and β, and subsequently in the likelihood ratio of the criterion value. The difference between the first and the second criterion can be interpreted as an increased tolerance towards false positive decisions: deciding to leave the port, given that the weather will actually be stormy. But it also involves a shift in the prior beliefs. In summary, the decision making model of testing allows for prior-belief considerations to play a role as well, albeit quite invisibly in the setting of the decision criterion. In the Bayesian model, prior beliefs and the likelihood ratio are separate. They separately and independently contribute to belief revision in the way prescribed by the theorem.

The sailor example can also be used to illustrate the third parallel between the Bayesian model and previously discussed theories of testing. This concerns the asymmetry between the "evidential value" of a confirming and a falsifying result from a given test. In the present example, the likelihood ratio in Bayes' theorem may be interpreted here as the reliability of the weather forecast. Suppose that the weather forecast correctly predicts a storm in 90% of cases, whereas it also erroneously predicts a storm in 30% of cases (i.e., when the weather actually turns out to be calm). The likelihood ratio is .90/.30 for the "storm" prediction. Let us further suppose, for simplicity, that the prior beliefs of the tester are indifferent: The chances are 50-50 of it becoming stormy. The tester relies only on the forecast. In this case, it can be calculated that the actual "storm" forecast leads to three times more confidence in the storm than in the calm weather hypothesis. However, if the forecast predicts calm weather, the tester should be seven times more confident in the calm weather hypothesis than in the storm hypothesis (see Appendix 1). Interestingly, it is the forecast that is less expected by the tester to be given in the broadcast anyway (calm weather) that will have the highest impact on his beliefs.

The asymmetry between the value of confirmations and falsifications in theories of testing was discussed earlier in reviewing the philosophical theories and the decision making framework. In the Bayesian analysis this asymmetry is basically similar to that which was demonstrated in those theories. Indeed, we remarked that searching for falsifications of a hypothesis (maximising the chance of a rejection) means, *ceteris paribus*, minimising its power to reject as compared to the power to confirm. The Bayesian approach will be rounded off with a short exploration of its applications in the study of everyday human hypothesis testing.

Bayesianism as a model for human hypothesis testing

Unlike the decision making framework, the Bayesian model has been used extensively in modelling hypothesis-testing behaviour. It also serves as the basis for the experimental laboratory tasks (Skov & Sherman, 1986; Slowiaczek et al., 1992) (see Chapter 5). The popularity of the Bayesian model in hypothesis-testing studies may be the result of its popularity in the domain of human *inductive* reasoning research. The use of Bayesianism in this area generated a rapidly growing new paradigm in psychology (Tversky & Kahneman, 1982).

The application of Bayes' theorem in this field is quite understandable because of its essentially inductive character. But this application has also reached the field of hypothesis testing. In typical Bayesian studies of hypothesis testing, the tests are presented to the participants as likelihood ratios and the prior probabilities are given. The participant in these studies is supposed to choose, from the range of possible tests, the one that is most suitable for testing a given hypothesis. By varying the properties of the proposed tests (by presenting, for example, tests as likelihood ratios that deviate more or less from 1), and observing the preferences of choice, the participant's testing strategy is inferred. The Bayesian model seems to be an adequate approach for testing behaviour in those cases where prior probabilities can be attributed to the hypotheses tested. Also, the test situation should have the character of a revision of belief. In contrast to the decision making model, the test is not performed in anticipation of an action. Indeed, the Bayesian approach assumes that one updates a prior belief in a continuous rather than a discrete process of hypothesis evaluation, where the output of one test serves as the input (the prior belief) for the next.

Nonetheless, the Bayesian model has been applied in situations that do not meet those characteristics. For example, in most Bayesian studies of testing, the prior probabilities are given to (and not chosen by) the tester. They are generally said to be "indifferent". That is, the prior probabilities of H_1 and H_2 are set to .50/.50. This allows the researcher to observe how tests are chosen when there is no *a priori* reason to believe more in H_1 than in H_2. Allowing for non-indifferent prior probabilities would complicate the interpretation of test choice. Indeed, the strategy of choosing the likelihoods may be based (partly) on the particular subjective prior probabilities a participant has in mind, and calculations become quite complex. By using only indifferent prior probabilities, however, an important characteristic of Bayesianism is overlooked: the attribution of (subjective) probabilities to the hypotheses before testing them. Also, this particular way of applying the Bayesian model to the study of testing behaviour confounds it with a decision making approach. Friedrich (1993) indeed reinterpreted some findings of Bayesian experimental hypothesis testing research in terms of decision making behaviour. An advantage of the Bayesian approach as compared to decision making models is that testing behaviour, i.e., searching information with regard to a hypothesis, is combined with processing this information to update belief in the hypothesis. Thus, in a Bayesian approach, not only can the information search strategy be described, but also the way a given test result is used in the revision of belief. This actual revision by participants may be "normatively" compared to what the theorem calculates it to be (Klayman, 1995).

Let us briefly examine some examples of the everyday testing situations mentioned in the former section on decision making models, and examine whether or not they could be adequately modelled by a Bayesian model. In the example of the defective television, the tester is not primarily trying to update his or her belief, but rather attempting to act in the most cost-reducing way. Predicting testing behaviour on the basis of a Bayesian analysis seems to be not very sensible in such a case. On the other hand, the archaeologist may well be described as a tester revising his or her belief, because the utilities of decision errors play almost no role. The archaeologist may have prior beliefs about the hypotheses. For instance, there may be a prior guess about the date of the object on the basis of where it was found. This guess may be revisable in the light of examinations of the object in the laboratory.

The last formal theory described in this chapter bears some similarity to the Bayesian approach, especially because of its inductive character. It asks, how much information we can get from a piece of evidence. It does not typically represent the hypothesis tester as a belief updater but, rather negatively, as an uncertainty reducer.

INFORMATION THEORY

Testing as reducing uncertainty

Information theory is a collective term for formal theories about information. According to information theory, a piece of evidence can be given a value indicating how much information it contains (Shannon, 1948). The theory is based on the communication metaphor. It assumes a source transmitting a message to a receiver. The amount of information transmitted is quantified according to the following measure (Coombs et al., 1970):

$$I(x) = -\log p(x) \qquad (2.3)$$

The unity of I is "bits" (the logarithm being taken to the base 2); x is the "message", but it can also be an observed event, or some evidence that correlates with the state of nature. As can be seen in the formula, the amount of event information is inversely proportional to its probability. Thus, I indicate how "unexpected" the event is. This is an intuitive principle. The more surprising an event is, the more we perceive it as "informative". The second principle that lies behind the definition of I is the requirement that pieces of information should be additive. That is, if event x_1 has been observed and, after it, event x_2, then the total amount of information should be the amount conveyed by x_1 plus the amount conveyed by x_2, given that x_1 is already "known" (see Coombs et al., 1970). The requirement of the second principle is satisfied by the properties of the logarithmic function.

The information-theoretical framework also provides a measure for the amount of information one can *expect* from a test or an experiment. Assume an experiment X with possible outcomes x_i: The measure aggregates all pieces of information on all possible outcomes and calculates how much information we can expect to obtain from this experiment. In the language of information theory, this measure is called "uncertainty". This term captures the intuition that our uncertainty about the results of an experiment determines the amount of information it will convey. In information theory, a test or an experiment is said to *reduce uncertainty*. This measure is:

$$U(X) = -\sum_{i} p(x_i) \cdot \log p(x_i) \qquad (2.4)$$

where X is the experiment and x_i the possible outcomes. A property of this measure is that it takes its maximum value when all possible outcomes are equally likely. This is the case in which we are most uncertain and the test can reduce our uncertainty to a maximum extent. Notice, however, the difference between the interpretation of the "information" measure and the measure of "uncertainty". Although the most informative experiment *a priori* is the one with maximally equally probable outcomes; i.e., the experiment where we have no expectations about its results, the result once obtained subsequently tells us more if it was less expected, in advance, than the other results. Interestingly, the measure of uncertainty reflects the idea that an experiment reduces more uncertainty if it is less "biased" *a priori*, i.e., if the specific outcomes are less predictable.

In hypothesis-testing situations, a piece of evidence is interesting to the extent that it tells something about a hypothesis. In the previous formulas of information theory, the hypothesis did not appear. However, there is a measure, in the information-theoretical framework, which calculates the amount of information a piece of evidence can provide to support a given hypothesis. Actually, it was Fisher (1950) who developed this measure, which can be seen as a special case of measure (2.3). This is the log-likelihood ratio (see also Kullback, 1959) which is to be interpreted as the information value of a result in supporting H_1 against H_2. We assume a situation with two mutually exclusive and exhaustive hypotheses. Savage (1954) calls this measure "differential information":

$$DI(x) = \log \frac{p(x \mid H_1)}{p(x \mid H_2)} \tag{2.5}$$

This function has some interesting properties. First, its value is 0 whenever the likelihood ratio is 1. The observation tells nothing that can be used to differentiate between the hypotheses. This is consistent with Bayesian testing theory. Second, if the ratio is higher than 1, the value is positive. If it is smaller than 1 it becomes negative, indicating that an observation's being more frequent under H_1 than under H_2 supports rather than falsifies H_1 and vice versa. Third, it can be deduced from the function under whose conditions, regarding the likelihood ratio, the information value of a piece of datum (irrespective of its being confirming or falsifying) is higher than another result from the same test. Imagine the situation with two hypotheses H_1 and H_2, and a test X with two possible observations x_1 and x_2. In the assumed testing situation, $p(x_2 | H_1)/p(x_2 | H_2)$ is equal to $(1-p(x_1 | H_1))/(1-p(x_1 | H_2))$. In Figure 2.1, the values $DI(x_1)$ and $DI(x_2)$ are plotted for several tests, i.e., several values of the likelihood ratio of x_1, as an illustration.

In the figure, $p(x_1 | H_1)$ is kept constant (at .80). The curves represent the DI(x) values for all possible values of $p(x_1 | H_2)$ (set out on the horizontal axis) and thus for the resulting likelihood ratios $p(x_1 | H_1)/p(x_1 | H_2)$. Now, the differential information values of both outcomes are equal at the point

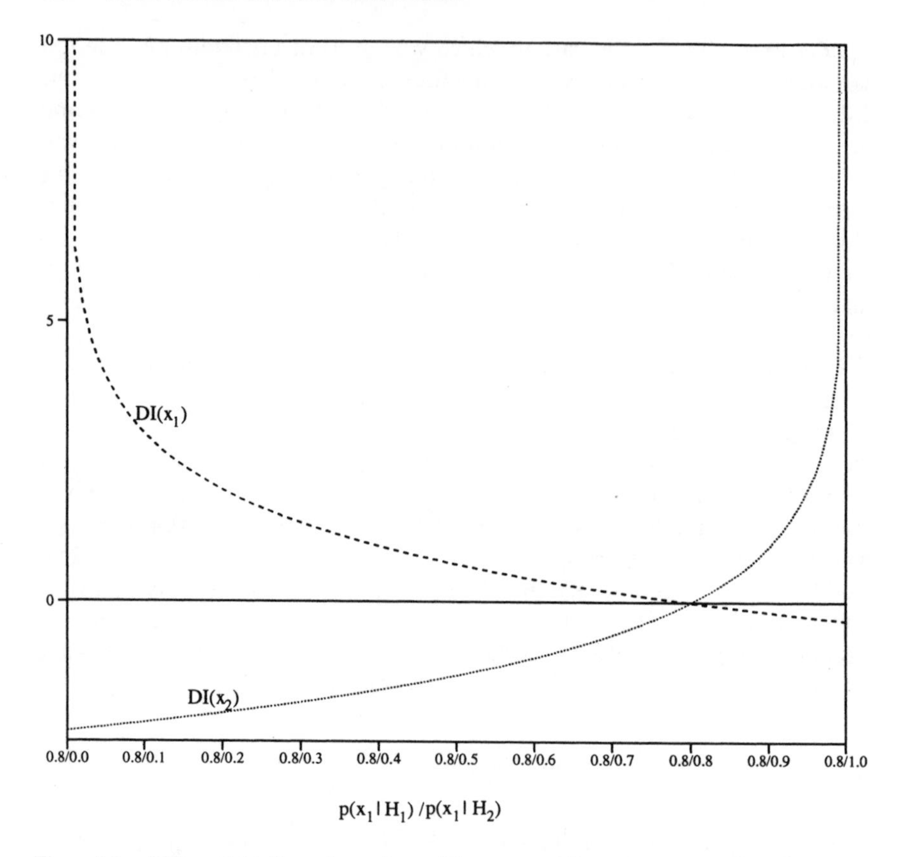

Figure 2.1. Differential information values of the two possible outcomes x_1 and x_2 of a test, as a function of the likelihood ratio of x_1.

where the likelihoods add up to 1. This is at the point $p(x_1 | H_1) = .80$ and $p(x_1 | H_2) = .20$, in the figure. All other tests, having other likelihoods, produce asymmetrical confirmations and falsifications. Either the confirmation can better confirm than the falsification can falsify, or vice versa. This asymmetry increases, as one of the likelihoods—either $p(x_1 | H_1)$ or $p(x_1 | H_2$—is more extreme (approximates either 0 or 1) than the other (being close to .50) as can be seen in the Figure 2.1 (see also Slowiaczek et al., 1992). This characteristic is relevant because in psychological studies of testing it has been found that humans tend to prefer extreme likelihoods for their focal hypotheses, and not-extreme ones for the alternatives. Hence, they tend to prefer tests with asymmetrical evidential values of outcomes. This phenomenon and its consequences are discussed in Chapter 5.

The *expected* information from a test to support one hypothesis against the other can also be calculated by means of the log-likelihood ratio, in the same way as the expected information of a piece of evidence was calculated. Imagine a test X with two hypotheses and the possible results x_i.

$$EDI(X) = \sum_i p(x_i \mid H_1) \cdot \log \frac{p(x_i \mid H_1)}{p(x_i \mid H_2)} \qquad (2.6)$$

This value is always positive. It decreases as the likelihood ratios approach 1, and it increases up to infinity as the likelihood ratio deviates from 1. The latter are the tests from which the highest amount of uncertainty reduction can be expected.

Some conceptual parallels between the information-theoretical principles and the mathematical and philosophical theories mentioned have already appeared in the present discussion of information theory. They will be further commented upon later.

Relation to other theories of testing

The basis of the information-theoretical approach is that a given observation is more informative when it is more surprising. This idea is consistent with the other theories on testing. In Popper and Carnap's theory, the test result was more severe or more relevant when it was less frequent *a priori*. An interesting contribution of information theory to the hypothesis-testing situation is the log-likelihood ratio information value and the expected log-likelihood ratio information of a test for a hypothesis, because it regards evidence in relation to a hypothesis. The likelihood ratio plays a prominent role in Bayes' theory as well. In line with Bayesianism, the information-theoretical measures calculate the amount of actual information or impact a result has, or the amount we can expect from a test. In both theories, variations of the likelihood ratio have the same meaning: The test is basically worthless when it approaches 1, for example.

In information theory, we assume that there is no prior knowledge about the hypotheses. The same "asymmetrical" property can be demonstrated, as we have already shown in the previous sections. This asymmetry also emerges in Popper's theory, and it has been shown to hold for the Bayesian revision of belief and decision making approaches as well. As a conclusion, many similarities can be found between the formal theories of testing. The similarities become clearer as we disregard the original philosophies, and focus on what they imply more concretely. In the final section of this chapter, I shall summarise and take a closer look at these similarities. I shall also return to the relevance the theories have for understanding hypothesis-testing behaviour.

IMPLICATIONS FOR TESTING BEHAVIOUR

From mathematical terms to dimensions of behaviour

All mathematical theories make direct or indirect use of the likelihoods of pieces of evidence. By means of these likelihoods, the test is described in a formal way. In decision making theories, they are used to calculate decision errors with regard to the truth of theories. In the Bayesian framework, they represent the impact of the evidence in the revision of belief in a hypothesis, while in the information theory they are the basis on which the amount of differential information of a piece of evidence is calculated. The consequences of the values of these likelihoods are globally similar in all formal theories, even if they differ philosophically. For example, if the likelihoods of a piece of evidence turning up are equal under both the theory tested and its rival, we will make as many wrong decisions as correct ones; the evidence will have no impact on our belief revision and it will not reduce any uncertainty. These are different descriptions for the same, quite inefficient, pragmatic behaviour of an everyday tester.

Some measures discussed previouly rest on the property of test outcomes, others on the prior quality of the test. The expected log-likelihood ratio is an example of a measure that says something about a test. The difference between this property of a test and the information a specific piece of evidence actually provides *afterwards*, is important for psychology, although in psychology, what is to be expected from a test is sometimes confused with what it conveys (Poletiek, 1996). Two tests having the same diagnosticity in the present definition, may each have a different distribution of the values of the possible outcomes. In practice, a human hypothesis tester may be concerned with the total expected amount of information *a priori*, regardless of its "direction". This might be called a symmetrical testing strategy (as we will also see in Chapter 4). But the tester may also be interested in the value of one of the specific possible outcomes that proves one hypothesis to be true (see Chapter 6). This is analogously called "asymmetrical" testing behaviour. Thus, the considerations resulting in a test choice may refer to either the total body of expected information or the values of specific outcomes. However, once a test with certain properties is chosen, the tester is "bound" to a precise interpretation of the result, which can be calculated by means of formal testing theories. The actual interpretation of the evidence in relation to test choice has recently been investigated in experimental studies (Slowiaczek et al., 1992). This is discussed in the following chapters.

What is a good test?

The statistical theories may be seen as containing a normative claim. Most of the authors of those theories have presented some ideas about how a test should be chosen. For example, Fisher proposed that the rejection area should be small. In science it should be put at .05. Along the same lines, Popper made a strong plea for "severe" tests. Neyman and Pearson produced a lemma that reduces the number of decision errors. They based their theories on the idea that the tester wants to reduce the number of decision errors of one kind, given that one tolerates a fixed number of the other kind. However, this theory essentially allows the tester to introduce pragmatic goals in the choice of a test. These goals are even reflected in the values of the parameters. In Bayes' framework and in information theory, the idea is common that tests with likelihood ratios deviating from 1 are "good".

With regard to the normative quality of formal theories of testing, it is important to notice that these pieces of advice to the tester (such as "maximise the likelihood ratio", "maximise the revision of belief", and "set the rejection area at .05") are not part of the formal working out of these theories. They are basically extra-mathematical. The fact that utilities can play a role in the Neyman–Pearson procedure does not tell one what they should be, and the fact that the uncertainty of a test can be calculated does not indicate that we should choose the largest uncertainty reducer among the tests. Therefore, there is no *formal* prescription to maximise any input term of these theories. Hence, according to the present argument, there is no sharp opposition between pragmatics and rationality. The pragmatic aspects of an everyday test situation, like goals and beliefs on the one hand, and formal rationality on the other, are not incompatible, but combined in these formal theories. When modelling human hypothesis testing with formal theories of testing, the task is to find out the parameters reflecting the personal utilities and beliefs people use as input, when choosing among tests. However, even when these parameters are found, the model might not be an adequate description of what people do. The divergence between such a description and actual test behaviour can have multiple causes: The (formal) theory might not be adequate for the testing situation at hand, the parameters reflecting people's goals might not have been correctly estimated, or the parameters are psychologically adequate, as well as the model, but the "mental" calculations do not lead to the same result as the formal model predicts. As Klayman (1995) puts it, there is no point in describing a test choice as "biased". The parameters are free. At most, it is the interpretation of the actual test result, given the quality and the parameter values of a once chosen test, which might be biased.

SUMMARY AND CONCLUSIONS

The formal theories all represent the testing situation differently, but basically they have much in common. Interestingly for the study of testing behaviour, the theories often leave room for subjective choices as to how a tester designs a test situation. Apparently, there is hardly a coherent theory of testing thinkable that considers testing as a procedure whose outcome is uniquely determined by the "facts"—not even a hard-boiled mathematical theory. This is also intuitive; if we want to gather information about some conjecture, there are generally many ways to define the relevant information we want to check. Our preferences for this or that strategy can be induced by prior beliefs or pragmatic utilities, or by our specific interpretation of the hypothesis.

There are two important features of performing tests according to the hypothesis testing theories of the present chapter and the foregoing one. These features may not have received the attention they deserve in psychological models of testing behaviour. First, all theories account for the *values* of test results of a given test. Choosing a test is not only anticipating how likely a confirmation is but also how much it tells about the hypothesis. The second feature is that test results of one and the same test can differ with regard to what they tell. This implies that one test can be a good possible falsifier and another a good confirmer of our idea. Considering these features can be an important aspect of testing behaviour.

In the forst two chapters, I have pointed to a number of parameters that play a role in test choice according to the theories of testing. Finding out how test choice is performed is a task for psychology. On the one hand, it can describe, by means of these theories, how humans proceed: How severe is the test choice? Under which conditions? How much uncertainty is reduced? And so on. On the other hand, psychologists have reported deviations from the normative claims of these theories. Testing is then described as biased with regard to one of the theories. The most salient finding is that human testers are liable to a confirmation bias. This finding has had an enormous impact in cognitive psychological literature. Although this conclusion might seem quite straightforward, it is far from that. One problem is that the theories are not easily translatable in a normative model for behaviour. Also, another standard than the one used may be equally adequate to analyse the testing behaviour. A very striking example of this problem is our demonstration that falsificatory test choice boils down precisely to confirmatory test choice. Ultimately, this would imply that confirmation bias is simply an impossibility. Indeed, tending to confirmation is necessarily tending to falsification.

In the following chapters, we turn to the psychological studies of testing behaviour. The philosophical and formal models will be used as a basis for understanding and interpreting the empirical work, especially the findings on confirmation bias.

CHAPTER THREE

Wason's rule discovery task

INTRODUCTION: WASON'S EXPERIMENTS ON HYPOTHESIS-TESTING BEHAVIOUR

The philosophy and the logic of testing theories has been outlined in the previous chapters. In this and the following chapters, the psychology of testing will be discussed. In a hypothesis-testing experiment, a tester is typically assigned a task in which hypotheses are to be tested and sometimes evaluated. This can be done in various ways. Every possible way is a test strategy. Thus, the researcher invokes a number of test strategies among which the participant makes a choice. From this choice, his or her testing behaviour is inferred. Wason designed two hypothesis-testing tasks, which became standard experiments. Test selection is the central issue in Wason's (1966) "selection task", as we shall see in the next chapter. In this task, the hypothesis is given by the experimenter, and the behaviour in question is the test choice. The whole process, starting with test choice, through hypothesis revision to the final evaluation of hypotheses, is studied in Wason's (1960) "2–4–6" task, which is discussed in the present chapter. In contrast to the selection task, the 2–4–6 task, occasionally called the "rule discovery" task, asks the tester to discover a "true" rule. This can be done by generating and testing hypotheses a number of times, subsequently evaluating and, if necessary, modifying the hypotheses on the basis of the test results.

Wason's two tasks have been the vehicle of almost all the "thinking" about hypothesis-testing behaviour among cognitive psychologists. Many theories of testing, even conflicting ones, are supported by some variation on one or other of these tasks. The domination of these two tasks in the study of hypothesis testing has the advantage of making the pieces of research easily comparable. Manipulations and their effects can be mutually compared, allowing for a coherent and increasing body of knowledge on how people perform tests. The disadvantage is, of course, the limited diversity of the task situations, and therefore poor correspondence to real testing situations. Because of the central role Wason's tasks occupy in the literature and in theorising about testing behaviour, however, two chapters are devoted to these tasks. The recent debates, theoretical explanations, models, claims, and the controversies they have generated are discussed.

WASON'S RULE DISCOVERY TASK, OR 2–4–6 TASK

Wason's participants had the job of discovering a rule governing a combination of three numbers (Wason, 1960). These testers were first presented with an example consisting of three numbers which conform to the rule, without the rule actually being made explicit. The true rule was simple: "Three numbers in increasing order of sequence". The example given to the participants was "2–4–6". In compliance with Wason's intention, almost all testers developed the hypothesis, based on the example, that the rule is: "Three *even* numbers in increasing order of sequence".

The task consisted of testing hypotheses until the rule had been discovered. A test consisted of the tester displaying three numbers to the researcher who then responded "yes" or "right" if the triple conformed to the correct rule and "no" or "wrong" if it did not. When the tester announced that he or she had discovered the rule, the experimenter stated whether or not the proposed rule was correct. The tester would then continue with the task, generating more triples and announcing more hypotheses, or stop if the rule had indeed been discovered.

Wason's original analysis of the task

Wason distinguished between a confirming and a falsifying strategy. The confirming strategy consists of proffering combinations of three numbers to the experimenter, which comply with the tester's own hypothesis concerning the rule, or, to put it another way, combinations that are examples of the tester's hypothesis. The falsifying strategy is defined as testing by means of combinations that do not comply with the hypothesis, thus, counter-examples. According to Wason, the use of hypothesis-compatible triples indicates "enumerative" thinking, while the falsifying strategy is evidence of "eliminative" thinking.

Wason's approach is normative. By means of experiment, he wishes to observe whether or not testers perform tests rationally. In this experiment, performing the test well means, for Wason, applying an eliminative strategy. The rationality of the strategy seems to be supported empirically by Wason's findings, in the sense that it appears to be effective in discovering the rule. Wason calculated the ratio of the number of eliminative to the number of enumerative attempts for each tester. Testers who were able to announce the correct rule first time had, on average, a significantly higher "eliminative-enumerative index" (1.79, n = 6): They used a falsifying strategy more often than a confirming one. Testers who first announced an incorrect rule as the answer had an index of 0.24 on average (n = 22). Enumerative testing has become known as "confirmation bias". Wason's conclusion is (1960, p. 139):

> The results show that very few intelligent young adults spontaneously test their beliefs . . . .

Notice that Wason identifies "testing" with "eliminative testing" here. This reminds one of Popper, Wason's source of inspiration, who alternatively spoke of falsifiability and testability. Popper's influence on Wason is also recognisable in the analogy Wason makes between the laboratory task and the scientific testing situation (1960, p. 139):

> The task simulates a miniature scientific problem in which the variables are unknown, and in which evidence has to be systematically adduced to refute or support hypotheses. Generating an instance corresponds to doing an experiment, knowledge that the instance conforms, or does not conform, corresponds to its result and an incorrect announcement corresponds to an inference from uncontrolled data.

Wason's experiment has been the impulse for an experimental and theoretical programme in cognitive psychology which has not ceased up to the present day. The first replications were directed towards Wason's operationalisation of the "eliminative strategy" and the confirmation bias. They are discussed in a later section. Subsequently, the influence of a number of task variations on discovery performance is discussed. Next, explanatory models of hypothesis-testing behaviour are dealt with, including Wason's original Popperian explanation and a new Popperian application. Finally, a few alternative models are reviewed.

Positive testing and confirmation bias

The first reaction to Wason's experiment came from Wetherick (1962). Wetherick discusses two aspects of the experiment. Primarily, he indicates that Wason's testers cannot refute the induced hypothesis by generating compatible test items. The deception stems from the fact that all triples that

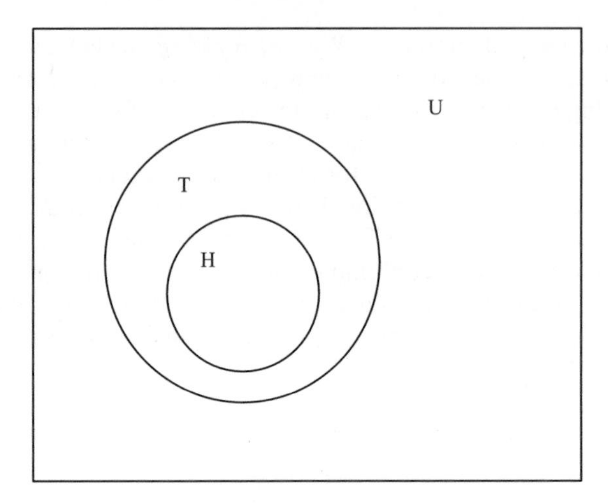

T: set of triples conforming to the rule
H: set of triples conforming to the hypothesis
U: set of all possible triples

Figure 3.1. The relation between the participants' hypothesis and the true rule in Wason's rule discovery task (1960).

fulfil the tester's induced hypothesis also fulfil the true rule. This argument is crucial and was actually only fully worked out years later, by Klayman and Ha (1987). The issue is as follows: Due to the relationship between the tester's hypothesis and the rule in Wason's formulation, all triples that fulfil the tester's hypothesis also conform to the true rule. If the tester now tests her hypothesis against items that fulfil this hypothesis, she will never receive a refutation but always a confirmation of the true hypothesis. The chance of refutation is zero. This is illustrated in the Venn diagrams of Figure 3.1.

Wason's formulation always leads to confirmation when one tests "enumeratively". This confirmation is subsequently interpreted by the tester as being evidence for the correctness of the hypothesis. This hypothesis was, however, incorrect, as the figure illustrates, and the reason that it is incorrect is because it is too restrictive. But this restrictive hypothesis was induced precisely by the experimental stimuli. This situation, created by Wason's experimental design, need not, of course, represent the general case. For example, one may also have a hypothesis that is too "loose", or a hypothesis that overlaps the correct rule. Thus, in Wason's specific design, the enumerative strategy leads to erroneous answers to the task, but is not erroneous universally. This argument has led Wetherick to refer to the enumerative test strategy as merely a "positive strategy". Testing by means of opposite examples—the eliminative strategy—is called the "negative" test strategy. These terms are also adopted here.

Wetherick's second point of analysis is linked to the first. This concerns Wason's presumption of having shown that some testers really do perform the task in an eliminative manner. Wetherick contests the view that Wason is able to show that "eliminating" testers *aim* to refute their hypotheses in the experiment. Wetherick argues that these testers can only be ascribed a real falsifying intention in the following two cases: (1) when they present a test item that is an *opposite example* of their hypothesis and which they expect to *conform* to the correct rule; and (2), when they present a test item that is an *example* of their hypothesis and which they expect *not to conform* to the correct rule. This then reveals that they are interested in refuting their hypothesis by means of such a strategy. When they present a test item that does not conform to their hypothesis and which they also do not expect to conform to the rule, they are not working eliminatively but are rather seeking confirmation of their hypothesis.

Wetherick designed a replication that illustrates his point. He asked testers to give their estimate, each time, of whether the test item, by means of which they wished to test their hypothesis, would or would not conform to the rule. In this way, he could check whether or not people do indeed attempt to falsify their hypotheses. Testers who do actually behave in a falsifying manner present either positive items that will not conform to the rule in their estimate (they expect "no" from the researcher), or negative items that do conform to the rule in their estimate (they expect "yes" from the researcher). Wetherick discovered that testers carry out this sort of testing relatively seldom. Of the successful testers who suggested the correct rule first time, an average of only 1.5 of the tests (mean number of tests performed was 9.15) were performed in a falsifying manner as defined by Wetherick. In short, eliminative tests were seldom consciously applied, and when they were applied this was not exclusive to the successful testers. This is a clear contradiction of Wason's claim that those who quickly discovered the rule were mainly geared to the falsification of their hypotheses.

However, the problem with falsification in this task is even more complicated than Wetherick demonstrated. Indeed, are participants able to conduct this normative strategy in the first place? That is, do participants feel that they *can* force the falsification of their hypotheses by strategic test choice only? In a study of my own (Poletiek, 1996), I manipulated the rule discovery task further, in line with Wetherick's experiments. Participants were explicitly instructed either to choose test triads in such a way as to confirm their hypothesis about the rule or to choose them in such a way as to falsify their hypothesis. Participants instructed to falsify preferred negative tests. However, almost all participants in this condition expected their attempt to falsify to fail. That is, they expected that the negative test triads would not conform to the true rule, and that the positive test items would. This is equivalent to expecting the chosen test to result in a confirmation, as Wetherick had demonstrated. Thus, my participants indicated that they were unable to bring about a falsification by their testing strategy alone.

These findings challenge Wason's conclusion that testers who use negative tests aim at falsification. They also challenge Wetherick's conclusion with regard to people's capability to perform tests that are expected to falsify their hypotheses. Wetherick shows that few people spontaneously *use* a strategy directed at falsification. Poletiek showed that people felt *unable* to employ a strategy that leads to falsification. Indeed, even testers clearly intending to refute their guesses by means of the test felt incapable of succeeding. In Poletiek (1996), the following argument is proposed to explain this incapability. The tester gives his or her "best guess" of the hypothesis and accordingly expects, by implication, that this will be confirmed by the test, regardless of the kind of test performed. Conversely, knowing which triples will actually lead to a rejection of the hypothesis implies that the hypothesis is not the "best guess". Indeed, some knowledge must tell the tester which facts there are in the world that conflict with the hypothesis. Falsification in the rule discovery task seems a question of the relation between the hypothesis and the true situation, and not just a matter of choosing a test. This is what participants also think.

Although the previous findings show that positive testing is not necessarily a "confirmation bias", in many studies with the rule discovery task a preference for positive testing is found, and is interpreted as non-normative in some way (Gorman & Gorman, 1984; Gorman, Gorman, Latta, & Cunningham, 1984; Gorman, Stafford, & Gorman, 1987; Kareev & Halberstadt, 1993; Kareev, Halberstadt, & Shafir, 1993; Mahoney, 1976; Mynatt, Doherty, & Tweney, 1977, 1978; Tweney et al., 1980; Wason & Johnson-Laird, 1972).

In a number of replications of the rule discovery task, attempts have been made to encourage the testers to apply a falsifying or "disconfirming" strategy (Gorman & Gorman, 1984; Gorman et al., 1984, 1987; Kareev et al., 1993; Mynatt et al., 1977, 1978; Tweney et al., 1980). This is referred to as "debiasing". In most of these experiments, the unbiased strategy is defined as "negative testing". In fact, the testers were instructed to apply a negative test strategy: They were encouraged to present triples that were opposite examples of their hypothesis. Tweney et al. (1980), for example, replicated Wason's experiment exactly. However, they divided the testers between two conditions. The first group received the "confirming" instruction, the second group the "disconfirming" instruction. Gorman and Gorman (1984) and Gorman et al. (1987) used the same instruction. In Tweney et al.'s instruction, the experimenter specifies that a "confirming" test must be able to conform to the rule. A "disconfirming" test is a test item that one expects not to conform to the rule. Interestingly, this is exactly the opposite of Wetherick's (1962) and Poletiek's (1996) argument that defines a disconfirming strategy as using those negative tests that one in fact expects to conform to the rule.

The "disconfirming" instructions appear to influence primarily the test strategy, and sometimes the discovery of the rule. In the studies performed by Tweney et al., Gorman and Gorman, and Gorman et al., the testers applied twice as many negative tests in the "disconfirming" condition (where testers were instructed to apply a negative test) as in the confirming and the control conditions. These negative tests led to actual refutations when the relation between the true rule and the hypothesis was similar to the one originally used by Wason (1960). Kareev et al. (1993) induced negative testing by presenting participants both with triples proposed by the experimenter and with test results. Having been given the opportunity to acquire some knowledge about the relation between the true rule and the hypothesis in this way, participants saw the usefulness of negative tests in this task. Kareev and Halberstadt (1993) conclude that when participants know the relation between the true rule and the initial hypothesis, they prefer negative tests that lead to rejection of their hypothesis. Thus, the "debiasing experiments" show that negative testing does not *always* lead to *informative* refutations; this occurs only when the true state of affairs allows it. Kareev et al. showed the influence, on negative testing, of having some knowledge about the relation between the hypothesis one has and the true state of affairs. When participants have insufficient knowledge about this relation, they feel unable to evoke falsifications by adapting their strategy, which is congruent with the logical analysis of the task structure (Klayman & Ha, 1987; McDonald, 1990, 1992; Poletiek, 1996).

Another way of looking at positive and negative testing is to regard them as *sufficiency* and *necessity tests*, respectively (Klayman & Ha, 1987; Spellman, López, & Smith, 1999). This describes the strategies in terms of the kind of information they provide. Thus, testing positive items, that is, items that conform to the hypothesis, provides information about whether the conditions formulated in the hypothesis are sufficient (if the item conforms to the rule) or not (if the item does not conform to the rule). In other words, positive tests only allow for discovering false positives. However, neither test outcome tells whether or not the hypothesis is necessary to predict the rule. Negative tests give the latter kind of information. They make it possible to discover false negatives. In other words, testing with sufficiency tests informs the tester whether or not more restrictions should be put on the hypothesis; it does not inform about whether or not the hypothesis has *too many* restrictions. One might put it as follows: The bias resulting from positive testing is not a "confirmation" bias but a "restrictiveness" bias. This is indeed what Klayman and Ha (1989) and Spellman et al. (1999) found. In their rule discovery experiments, participants tended to perform positive tests that led them to over-restricted hypotheses. They sometimes do so precisely with this purpose in mind (Spellman et al., 1999).

Another factor influencing the test result is also discussed by Klayman and Ha (1987). They show that the probability of obtaining falsifications is dependent on the "base rate" of the items satisfying the true rule, and of the items satisfying the hypothesis. When these base rates are small and the hypothesis is reasonably "close" to the true rule, positive testing is much more likely to provide falsifications than negative testing. This is also intuitive. If we have a hypothesis about a "small phenomenon", the set of negative test items is huge and the probability of finding a "true" item among them is quite small. It is better, then, to look at hypothesised items if we want to find one that is not true. Klayman and Ha (1987, 1989) explain preferences for positive testing by assuming that, on the one hand, hypotheses are generally about rare phenomena and are a reasonable approximation of the truth. The tester makes this assumption in the laboratory task on the basis of real-life experience. On the other hand, testers are, in actual practice, generally more concerned with false positives than with false negatives. Both assumptions lead to the preference for positive tests, which is a rational preference on the basis of Klayman and Ha's arguments. But the proof of the pudding is in the eating. A testing strategy may be rational, but does it make the task of finding the true rule easier? What are the determinants of discovery performance? This question is addressed in the following section.

Determinants of success in the rule discovery task

Do refutations help? Does one discover the rule more quickly when one *receives* falsifications more often? This is the Popperian presumption, discussed in Chapter 1, which is the foundation of the idea of confirmation bias as a bias: It is not effective to confirm; one learns more from refutations. There are some studies that support this. Gorman et al. (1987) claim that testers who were successful in their "disconfirmation" discovered the rule more quickly, although the facilitation did not work with all "true rules". However, this finding is disputable owing to the specific definition of the "successful falsifier" (Gorman & Gorman, 1984; Gorman et al., 1987). According to Gorman et al., testers receive a falsification when they follow a negative test strategy, when they expect "no" as a result, and also when they receive a "no" from the researcher. However, this combination actually leads to a(n) (equivocal) verification, according to Wetherick (1962) and Poletiek (1996). Logically, only a positive test with a "no" and a negative test with a "yes" should make the tester reject a hypothesis, as was argued in the previous section. Thus, at this point, it is not unambiguously shown that disconfirming helps to find the truth.

Since Wason's participants were originally unsuccessful because of the numerous ambiguous confirmations they received, one would expect that participants who actually receive falsifications and accordingly reject the falsified hypotheses would perform better. Klayman and Ha (1989) varied the relation between the true rule and the participant's initial hypothesis. In one condition, the hypothesis was broader than the true rule. In another condition, the hypothesis was embedded in the true rule (as in Wason, 1960). In the last condition, the hypothesis and true rule overlapped. As predicted, testers using positive tests got more falsifications in the surrounding condition than in the embedded condition. However, they did not discover the rule more often than did the testers in the other conditions. Thus, actually getting refutations *per se* did not help in finding the truth. Poletiek's (1996) finding supported this conclusion: A negative relation was found between the number of falsifications obtained and the discovery of the rule. In particular, in cases when the rule was very difficult to find and therefore the participant's belief in the hypotheses they generated was very low, any falsifications seemed more of a hindrance than a help to the discovery of the true rule. Interestingly, in complex hypothesis-testing tasks, testers may even neglect refutations (Mynatt et al., 1978). Neglecting negative test results often led to successful discovery in this experiment. The irrational—from a normative point of view—strategy was actually effective in practice.

The findings of Klayman and Ha (1989) and Poletiek (1996), together with several other replications of the rule discovery task, suggest that testers need to have some minimal global knowledge about what the true rule looks like in order to recognise the value of falsifications and to actively look for them (Gorman, 1989; Tweney, Doherty, & Mynatt, 1981). This two-stage process of hypothesis testing consists of starting off by focusing on the evidence that is consistent with the hypothesis and then, after having developed a hypothesis with some reasonable probability, trying to falsify it. In Poletiek (1996), participants in a condition without any prior knowledge were not successful in finding the rule because they could not develop a reasonably likely hypothesis within the limited number of trials they had. The falsifications they obtained were of no use to them because the possible alternative revisions after such a falsification were almost infinite. As testers have (and take) the opportunity to perform more tests, the likelihood of finding the correct rule increases (Klayman & Ha, 1989).

Mynatt et al. (1978) found that, in a very complex variation on the rule discovery task, participants focused on confirmation particularly in the early stages of the task. Mynatt et al. deduced this tendency from protocols of testers which were closely monitored. Tweney et al. (1981) and Klayman and Ha (1989) found that testers also tend to "test alternative hypotheses", which is a form of seeking falsifications and simultaneously confirming possible alternative hypotheses, by means of crucial evidence (see Chapter 2). This strategy was carried out mainly in the latter stages of the testing process.

The work of Halberstadt and Kareev (1995) and Kareev and Halberstadt (1993) revealed similar phenomena. These authors focus on the difference between a "reception mode" of testing hypotheses and a "generation mode". In the first mode, testers are presented by the experimenter with triples that are said to be "correct" or "incorrect". In the second mode, which is equivalent to the traditional methodology in the rule discovery task, the participants could test triples composed by themselves (Halberstadt & Kareev, 1995). The first mode was characterised as "observing and formulating a hypothesis". When participants were given a choice to use either the one or the other strategy, they chose both. However, they tended to work under the reception mode first and then to generate triples. In other words, participants first try to build a hypothesis and get some confidence in it before submitting it to tests that might result in falsification.

Another interesting finding in the study of Halberstadt and Kareev is that participants obtaining a great many confirmations of their tentative hypothesis during the first "hypothesis building" stage of the testing process tended to actively build experiments (propose triples) less often. Conversely, when few observations fitted the hypothesis they had in mind, participants tended to switch to the more active testing mode earlier. The fact that observing many confirmations led to less active testing activity may partly explain why participants in Wason's original task version failed to find the rule. These testers were exposed to many confirmations, which gradually increased their belief in the hypothesis to a point at which further testing was no longer felt to be necessary. They concluded that the hypothesis was the true rule. "Wason's participants spoke too soon", as Klayman and Ha (1989) expressed it.

From the foregoing studies, we might advance the following tentative conclusion. Confidence in the hypothesis may be related in a curvilinear way to the search for and use of falsifications. Indeed, when the tester is very uncertain about a hypothesis or when all possibilities are still open, he or she may be primarily interested in finding some observations that grant some plausibility to his or her idea. Subsequently, when the hypothesis has gained some plausibility (because a few confirmations have been observed), but is still competing with alternatives, counter-examples of it might be explored. When these falsifications are not found, the tester becomes very confident in the hypothesis, because he or she has observed a lot of only confirming evidence; the hypothesis is strongly confirmed. Hence, the hypothesis is accepted as true (as Wason's participants did) and the tendency to search and use falsifications decreases again. In some cases, as in Wason's experiment, the confirmatory outcomes are misleading, however, and performance is impeded, but this is not the general case. This curvilinear pattern also fits in with some findings of Klayman and Ha (1989) about "testing alternative hypotheses". Klayman and Ha found that when the rule discovery task is divided into three stages over time, few "alternative tests" are performed in the first stage, most are performed in the second stage, and in the last stage few again are carried out. More generally, when we develop hypotheses about the world, we are first oriented on what *fits* in with a preliminary hypotheses. Then, we check the limits of its validity. Finally, when no more serious exception emerges, we accept it and direct our attention again to the confirming evidence.

In line with this argument, Klayman and Ha suggest that time to build up a hypothesis, and to test it thoroughly, helps in the rule discovery task. Falsifying by means of generating and selecting plausible alternatives is recognised in most studies as an efficient strategy for discovering the true rule, at some stage in the discovering process. I turn to this strategy now.

Testing alternative hypotheses. One manipulation that causes a dramatic increase in performance in the rule discovery task is the famous DAX-MED variation (Tweney et al., 1980). The authors modified the task into a classification task. In this experiment, the researcher did not give the answer "right" when the test item sufficed—instead the answer was "DAX". When the test item did not suffice, the answer was "MED". The testers were instructed to discover two rules: the rule that generated DAX triads and the rule that generated MED triads. DAX was the original rule in Wason's task of three increasing numbers. The MED rule was "any other combination of three numbers". According to the authors, this variation should be able to meet the objection to the standard version of the task in which the researcher uses the words "right" and "wrong" or "yes" and "no". This original feedback formulation could lead to the tester being discouraged by the use of "disconfirming information" since this is indicated by the word "wrong" or "no". The meaningless words "DAX" and "MED" have no negative connotations, thus any influence of this connotation is eliminated.

The proposal of triads that conform to the hypothesis was originally called a confirming strategy. This means that when one tests either the DAX or the MED hypothesis by means of compatible triads this is called a confirming strategy. Accordingly, the confirming strategy simply corresponds to Wetherick's positive test strategy (1962). The effect of this semantic variation of the feedback appeared to be enormous: 60% of the testers presented the correct DAX rule as the first response, although most testers had applied a positive test strategy. In comparison with other replications, a "positive" test strategy was used more frequently, and a negative strategy relatively less frequently. The authors suggested that the procedure increased performance due to the fact that information that could merely be interpreted as "erroneous" in previous tasks may now be attributed to a rule.

The DAX-MED effect has been replicated several times (Gorman et al., 1987; Vallée-Tourangeau, Austin, & Rankin, 1995; Wharton, Cheng, &

Wickens, 1993). In general, considering more than one hypothesis and testing them simultaneously significantly facilitates discovery performance (Farris & Revlin, 1989; Klayman & Ha, 1989; McDonald, 1990; Tukey, 1986). The instruction to search for two hypotheses (dual goal) instead of one hypothesis (single goal) is apparently a crucial cognitive heuristic in the successful performance of the task. One explanation is that testers simply perform more tests when asked to find two rules (Wharton et al., 1993). Farris and Revlin (1989) and Klayman and Ha (1989) also found that performing more tests increases discovery performance. Another explanation is that the information from the tests is of a higher quality than that which is received in a single-goal testing situation (Vallée-Tourangeau et al., 1995; Wharton et al., 1993). I will shortly go into this debate.

A version of this second explanation is given by Evans (1989). Primarily, the presence of two hypotheses induces positive testing of one and negative testing of the other. Evans also agrees that a positive or neutral "labelling" of test outcomes (MED instead of "incorrect") has a favourable effect on reasoning. Wharton et al. (1993) argue, however, that the quality of the information obtained in the DAX-MED version is only better if the two rules are complementary. Indeed, consider the classical version of the 2–4–6 task. DAX is the original rule and MED the complement. A tester who tests DAX positively (sufficiency testing) will become informed about how to restrict the hypothesis but not about how to broaden it (Klayman & Ha, 1987). However, as the tester is stimulated to test the MED hypothesis as well, he or she will become informed about how to restrict the MED hypothesis. Each time the MED hypothesis needs to be restricted on the basis of a falsification, it is immediately implied that the DAX hypothesis needs to be broadened accordingly, since both are complementary. Thus, testing the complement, even positively, prevents narrowing down the "focal" hypothesis too much. And, as explained above, a major problem in Wason's task is that testers narrow down their hypothesis too much.

Wharton et al. (1993) tested the information quantity and the complementarity hypotheses against each other. They found that when testers did not know about the complementarity relation between the two rules (the instruction implied that the relationship could be overlapping, embedding, or of any another type), they were not better in discovering the rule than "single-goal" testers (instructed to test one hypothesis only). However, when the complementarity of both DAX and MED was explicitly mentioned, the "dual-goal" participants performed better. In addition, they showed that increasing the number of tests *per se* did not enhance performance: Indeed, single-goal testers who were encouraged to increase the number of tests did not perform better than control participants. Finally, Wharton et al. also criticise the explanation by Tweney et al. (1980) and Evans (1989) that participants make better use of information neutrally labelled "MED" than information negatively labelled "wrong" or "incorrect". Indeed, participants in the single-goal condition mostly announced their (incorrect) rule before obtaining any negatively labelled feedback.

The crucial role of knowing the complementarity of the two rules in order to improve performance was challenged, however, by Vallée-Tourangeau et al. (1995). They explicitly varied the relationship between the DAX and MED rule in their instructions (overlapping, embedded, independent, and complementary). They found no influence of this instruction on discovery performance. The only factor affecting performance was testing one or two hypotheses. However, Vallée-Tourangeau et al. also rejected the information-quantity explanation. They found that when keeping the number of tests constant, dual-goal testers still performed better.

The incongruence of the results obtained by Wharton et al. and those obtained by Vallée-Tourangeau et al. may be due to differences in instructions. This is a general problem with DAX-MED experiments: It is difficult to be sure whether or not participants actually mentally represent the two rules as complementary and exhaustive, on the basis of the instructions. In Tweney et al.'s original version, the DAX and MED rules were intended to be complementary but also to be an exhaustive description of the hypotheses space. That is, each possible triple was either DAX or MED. Those characteristics are more precise in Wharton's instruction ("Triples can be sorted into two separated groups which will be referred to as DAX triples and MED triples") than in Vallée-Tourangeau's instruction ("I have in mind two rules that specify how to make up sequences of three numbers . . . You should produce number triples and I will tell you whether they are DAX sequences or MED sequences"). The fact that Vallée-Tourangeau et al. found no effect of the complementarity instruction may be due to the fact that participants did not all actually interpret the relationship as complementary. This explanation is supported by the protocols of some of Vallée-Tourangeau's participants, who came up with DAX and MED rules that were not complementary. Interestingly, Vallée-Tourangeau et al. propose that the crucial factor explaining success was the "heterogeneity" of items proposed by the tester to test his successive hypotheses. Heterogeneity was measured by the extent to which an item could discriminate between several hypotheses. This heterogeneity is facilitated by the dual goal instruction, because participants had at least two hypotheses in mind simultaneously.

Thus, looking at the debate, two major recent explanations of the DAX-MED facilitation effect remain: The complementarity hypothesis and the test item "heterogeneity"-strategy the DAX-MED instruction induces. These two explanations may be less incompatible than the two positions suggest, when they are analysed against the background of the formal theories of testing. In Bayesianism and the decision perspective on testing, two or more complementary hypotheses are assumed, which also exhaustively describe the "hypothesis space", that is, the set of all possible hypotheses. After having specified the hypotheses, tests can be defined in terms of the likelihoods of possible observations under the assumption that one of the hypotheses is true. A test can be chosen on the basis of its power to discriminate between the two hypotheses, and the test result interpreted in terms of the increase in confidence in the hypothesis it supports. This might be a model of what happens in the DAX-MED situation. First, the tester starts by generating two hypotheses. This might automatically direct the tester's attention to the possible differences between them rather than to the possible overlap. In this manner a complementary representation is made. Thus, he or she may think of the test items that are probable under DAX and unlikely under MED, and vice versa, those that are probable under MED but not under DAX. Searching for those items means searching for the "limits" of DAX and MED ("heterogeneity" testing). In other words, the dual-goal strategy in hypothesis testing may stimulate diagnostic likelihood ratios. In the single-goal variation, only one hypothesis is considered. This means that an alternative is not clearly represented and the boundaries of the focal hypothesis may not be sought explicitly. More specifically, the probability of a test item being true under the assumption of the alternative hypothesis (the denominator in the likelihood ratio) is neglected. This is exactly the mistake participants tend to make in the original 2–4–6 task. Thus, looking at the DAX-MED facilitation by means of a formal Bayesian analysis can explain that this manipulation stimulates both a focus on the differences between the alternatives (complementarity) and on test items that discriminate between these alternatives (heterogeneity). Both aspects help and are related.

Apart from the DAX-MED experiments, the testing of alternatives has been investigated in other variations of the 2–4–6 task. In a number of experiments, the tester is instructed to generate more than one hypothesis (Green, 1990; Tukey, 1986; Tweney et al., 1980). In this, the strategy of testing alternative hypotheses is deliberately induced in the testers. In other experiments, the question is whether or not people pursue this strategy spontaneously (Farris & Revlin, 1989; Klayman & Ha, 1989). Tweney et al. (1980) gave testers the instruction to generate two hypotheses concerning the rule at each trial, and to present two corresponding test items. Green (1990) and Tukey (1986) encouraged the testers to generate as many hypotheses as possible at the beginning of the task. The testers could then proceed with testing in the usual way. In addition, Green instructed the testers to consider the relationship between the various hypotheses they had generated. Testers turned out to discover the rule more quickly if they "tested more alternatives" in the sense described above.

Klayman and Ha (1989) also found that use of "alternative tests" led more quickly to the discovery of the rule. They also observed that participants sometimes performed "limit testing". These were positive tests aimed at testing the limits of one hypothesis against a set of alternatives. Similarly, Oaksford and Chater (1994a) argue that people seek boundaries of their hypothesis in the process of testing. I shall return to this topic in the following section.

Farris and Revlin (1989) explored whether or not testers test alternative hypotheses by means of what they call "counter-factual reasoning". This test strategy consists of testing an alternative hypothesis B that one has in mind, by means of hypothesis A that one regards as being unsound. Notice that this is what happens in significance testing (Chapter 1). The authors found that testers who reasoned "counter-factually" discovered the rule more quickly. Oaksford and Chater (1994a) proposed an extension to this counter-factual strategy, which is psychologically less demanding, as they argue. Indeed, in Farris and Revlin's strategy, testers are supposed to infer a hypothesis from the example-triple. Afterwards, they generate a complementary hypothesis, which they test positively. When they get a "yes" from the experimenter, the complementary rule is considered as plausible, and this last hypothesis is taken as the starting point from which to repeat the process. Oaksford and Chater show that this model is logically not well justified, because the complementary of the original hypothesis (which is consistent with the example) is falsified from the very beginning. Apart from this illogicality, Oaksford and Chater propose that participants might indeed perform some counter-factual reasoning, but rather than generating repeatedly complementary hypotheses, people might reason on the basis of the evidence they get from the experimenter. Along this line of thought, people first generate a plausible hypothesis, then a hypothesis-inconsistent triple is proposed. If it gets a "yes" from the experimenter, the tester looks for common features of the example triple and the triple obtained. This is the basis of generating a new hypothesis. In this manner, the boundaries of the true rule are explored, but on the basis of concrete data. Alternatives are generated and immediately tested. Interestingly, Tweney et al. (1980) found that the instruction to generate many hypotheses, *per se* (without their having consequences for the test strategy), led to worse instead of better achievements. Thus when testing alternatives, people probably reduce as much as possible the number of hypotheses they consider simultaneously, while performing some comparative strategy (see also Mynatt, Doherty, & Dragan, 1993).

On the basis of those studies, it can be concluded that the test of alternatives almost always leads to good discovery performance in Wason's task. Its principal characteristic is the search for evidence that can separate two hypotheses, such as, for example, the hypothesis one thinks is most likely to be true and a "second-best" alternative. In the next section, I shall turn to the application of philosophical theories to the rule discovery situation. In particular, I shall discuss how Popper's severity principle, which can be regarded as a form of "alternatives testing", can be applied to the task, and what its benefits are. Finally, I shall present a few experiments that approach testing behaviour from the perspective of other philosophies.

Popperian and other philosophical standards applied in the rule discovery task

Wason (1960) originally intended to simulate a scientific reasoning situation in the rule discovery task. Would people spontaneously reason in the same way as Popper prescribed for scientists? Especially, do lay persons submit their ideas as much as possible to tests that may refute them? Wason's conclusions are stated in terms of people's capability and willingness to reject their hypotheses. Conservative testing behaviour, criticised by Popper, was operationalised as positive testing (confirmation bias), and normative Popperian behaviour as negative testing. We have seen, however, how this translation of Popper's ideas can be disputed.

Positive testing has been shown not to be conservative under all circumstances. Whenever the true situation is changed, the same behaviour can generate refutations (Klayman & Ha, 1987, 1989; McDonald, 1990), and apparently become normative in the Popperian sense. At this point, the question arises as to whether or not there is a way to operationalise "Popperian behaviour" in the rule discovery task at all. In the next section, I shall propose such an operationalisation. Test *severity* was measured in line with Popper's definition (formula 1.4) in the 2–4–6 task. As an illustration, this operationalisation is applied to some data of the 2–4–6 task in a secondary analysis.

Severe testing in the rule discovery task and its merits. Severe testing consists of testing observations that have a high probability of being true under the focal hypothesis and a low probability under all possible hypotheses. "All possible hypotheses" is referred to as the "background knowledge" in Popper's theory. Such an item in the rule discovery task would be one that has a reasonable probability of being true under the focal hypothesis, and also would be very unlikely to be true if we do not assume the focal hypothesis. A problem is that the number of "possible alternative hypotheses" is infinite (see also Oaksford & Chater, 1994a). Moreover, people cannot easily keep more than one hypothesis at a time in mind (Mynatt et al., 1993). However, we assume here that, in psychological terms, this set is not infinite. That is, in order to perform a more or less severe test, it is not necessary to generate explicitly *all* possible alternatives. Also, this set is "fuzzy", psychologically speaking. That is, the tester intending to perform a severe test will not always consciously formulate one by one all hypotheses she eliminates, yet be able to make a globally accurate estimation of its content and size. This reasoning is not necessarily explicit and precise.

We may assume that a tester knows, albeit vaguely, that a number of hypotheses can explain the example and, from this set, she takes the one that seems most likely to her: the best guess. Subsequently, a test item can be chosen that excludes a number (many or few) of this "fuzzy" set of alternative hypotheses. Excluding many of them is analogous to reducing the probability of the denominator of definition (1.4), that is, the probability of getting a "yes" answer from the experimenter if one of those alternatives is true. The probability of the numerator, of obtaining a "yes" answer assuming that the focal hypothesis is true, is 1, which reflects that it is a positive test. Thus the severity of a test is controlled by excluding more or fewer alternatives.

Interestingly, what is proposed here as a severe testing operationalisation in the rule discovery task approximates to what Oaksford and Chater (1994a) and Farris and Revlin (1989) call boundary testing, and Klayman and Ha (1989, p. 603) call "limit testing":

> Consider for example the participant who tested the hypothesis "ascending numbers" with the triple [–100,0,105] . . . . These are examples of what could be referred to as "limit testing" and we believe they represent deliberate attempts to maximise the chance of falsification if the hypothesis is incorrect, but within the positive testing strategy.

I investigated test severity in a secondary analysis of data from a rule discovery experiment (Experiment 2, participants in the conditions "one example" and instructed to "test"; in Poletiek, 1996). That is, I estimated the participants' severity of performed tests, and correlated this measure with the number of actual falsifications obtained and discovery performance. In this experiment, participants got "2–7–6" as an example of the rule. The true rule was "even–uneven–even number". The "severity" variable was scored for this group of 15 participants. The severity of a test can be equated with the number of alternative hypotheses one excludes. A test is more severe the more it:

- is a positive test and *excludes* more plausible alternative hypotheses, or
- is a negative test and *includes* more plausible alternatives.

Indeed, the numerator in the severity of test definition (1.4) ($p(x \mid H \text{ and } b)$) is 1 if the test is positive. The denominator ($p(x \mid b)$) decreases as the number of plausible alternatives (forming together the disjunctive set b) with which x is incompatible, increases. It is more difficult to think up a negative severe test than a positive severe test. In order to recognise the principle of the severe test in the case of negative testing, one can imagine that the *absence*

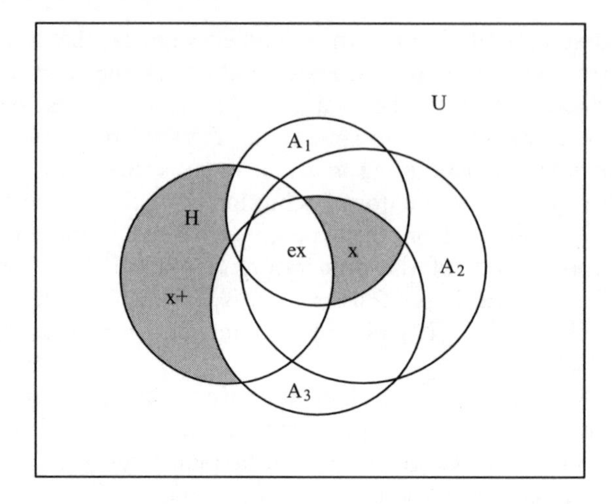

H: set of triples conforming to the hypothesis of the tester
A_1: set of triples conforming to the alternative hypothesis A_1
A_2: set of triples conforming to the alternative hypothesis A_2
A_3: set of triples conforming to the alternative hypothesis A_3
U: set of all possible triples
x+: set of severe positive triples
x : set of severe negative triples
ex: example triple

Figure 3.2. A severe positive and a severe negative test in Wason's rule discovery task.

of the observation, or a negative test result is predicted by the hypothesis but not by the plausible alternatives. This is shown in Figure 3.2. When one tests using a severe negative triple (x–), and this proves not to fit the rule (the test outcome is "no"), this is a strong confirmation of H. After all, all alternatives included in the test triple have been refuted in one swoop. A "plausible hypothesis" was defined as any hypothesis that was consistent with the given example. The whole principle can be illustrated by means of a Venn diagram (see Figure 3.2).

To make the calculations of severity, I took all plausible hypotheses generated by all participants to be the set of plausible alternatives (the background knowledge). As explained, its limits are not precise and are somewhat arbitrary. However, I believe that it sufficiently approximates the fuzzy set of possible alternatives that participants in this specific task situation have in mind. Moreover, I am not interested in the absolute magnitude of this set, but in differences between participants as to whether they exclude a smaller or larger share of the set of alternatives in their own testing. The collection did not turn out to be inordinately large and at first

TABLE 3.1
Set of plausible alternative hypotheses generated by the participants in the rule discovery task with example triple 2–7–6

1. Random series of numbers.
2. Three numbers under 10.
3. Three numbers of which the middle one is the highest.
4. Three numbers of which two are adjacent.
5. The first number is smaller than five, the second and third are larger than five.
6. The middle number is the largest, the first one is smaller than the middle one, and the third lies in between these two.
7. The first and last numbers are even.
8. Two even and one odd number.
9. Even–odd–even numbers.
10. Even–odd–even numbers, the middle one being the largest.
11. Three numbers of which the middle one is odd and the largest.
12. Even number, prime number, even number.
13. The second number is the first plus five, the third is the second minus one.
14. The difference between the first and the second number is five, between the second and third it is one.
15. The first and the third numbers are even, the middle number is odd, and the even numbers must be unalike.
16. 2, 6, an odd number.
17. The second number is larger than the first, and the third is the second minus one.
18. The first number is even, the second, odd and the third is the second minus one.
19. Three numbers which have no integers as square roots.
20. The second number is the first number times three, plus one, the third is the first number times three.

sight appeared to give a reasonably complete survey of the sort of hypotheses that were spontaneously evoked by the example. The set is shown in Table 3.1.

For each test executed, a calculation was made of the sum of the number of plausible alternatives from this list that are excluded if it is a positive test, and included if it is a negative test. This is added up for all the tests performed by one tester and divided by the number of tests that he or she has made (maximum of 5). This results in the average number of excluded hypotheses per test per participant. In this way, each tester received a severity score between 0 and 20 (each test could include or exclude a maximum of 20 hypotheses). This score indicates the mean severity with which a person tests his or her hypotheses in this task. The tests were scored according to severity by two assessors. Only 14 of the 15 testers were taken into consideration because one tester produced at least one implausible hypothesis at the beginning. The 14 testers performed a total of 51 tests. This produced 1020 judgements for each assessor. The inter-rater reliability was high: Cohen's kappa was .92. The average number of excluded alternatives in the tests was 8.90 in this group ($N = 14$; $s = 2.7$).

The first appraisal dealt with whether or not there was a relation between the number of excluded hypotheses and the number of falsifications obtained. For this, the proportion of obtained refutations was correlated with test severity. However, this analysis was only made with the subgroup that did not produce the correct hypothesis at the first attempt. After all, it cannot be expected of those in the group that found the correct hypothesis at the first attempt that they will obtain falsifications: An entirely correct hypothesis cannot be refuted. Six of the fourteen testers found the correct rule at the first attempt. If these six are not taken into consideration, the correlation between the number of falsifications and the severity of the test is high ($r = .82$; $p < .01$; $N = 8$). The more severe the test, the more often one obtained a falsification of the hypothesis. Thus pursuing falsifications *by means of test severity* indeed led to finding them.

Subsequently, the correlation between test severity and rule discovery was examined. In line with the Popperian idea, it was expected that the more one attempted to eliminate alternatives—thus, as one tested more severely—the better the discovery performance would be. Here also, in order to avoid the artefactual influence of immediate discovery, the group of six testers was omitted. Using the Mann-Whitney U test, the difference in test severity between the discoverers and the non-discoverers was gauged. It was not significant ($U = 3$; $p = .25$; $N = 8$). A final remarkable finding was that the negative tests were less severe than the positive. The correlation between the test type and the severity was $r = .66$ ($p < .01$; $N = 14$). Now, what can be concluded about the participants' truly Popperian attitude in the rule discovery task? In the foregoing analysis, the Popperian concept of severity of test was applied to the task. Testers appear to obtain falsifications more often when they test their hypotheses more severely. This is to be expected, since a severe test means that the tester makes the hypothesis take a large risk. A hypothesis that is not entirely correct can expect to receive a quick refutation with a severe test. The second finding is that conducting severe tests does not facilitate the discovery of the rule. This is also in line with what has been found in several rule discovery experiments: Obtaining falsifications is not a sufficient condition for success in discovering the truth. In contrast to falsifying testing as negative testing, test severity satisfies a "universality" principle. Basically, the tester can readily control the probability of getting falsifications in the future merely by manipulating the severity of his tests, without any knowledge about the relation between the hypothesis and the state of nature. He contrasts his best guess with more or less other possibilities he can think of.

The analysis performed above might be seen as an evaluation of Popper's prescriptive philosophy in the limited world of the rule discovery task. Indeed, Popper ordained that scientific testers should maximise the severity of tests. In the context of the rule discovery task, this standard may not always be helpful. This evaluation underlines the paradox in Popper's falsification prescription (see Chapter 1). Obeying the severity principle does indeed cause the occurrence of falsifications. However, actual refutations will hardly make the tester happy, since these refutations do not bring him or her nearer the truth. It seems that maximising the severity of tests or "limit testing" may be a sensible strategy when one has sufficient confidence that the hypothesis will survive these severe tests, because they can then provide strong support for it. However, starting off with severe testing right at the outset can end up with many rejections but no more knowledge. A reasonable alternative strategy to Popper's Spartan advice, is to start with a low-level "severe" test and progressively exclude more alternatives (see also McDonald, 1992), rather than maximising severity *per se*. This fits in with the argument presented above that testing a hypothesis starts mildly, becoming more severe as there is more evidence, and becoming mild again when one is quite sure about its validity. Following from the demonstration in Chapter 1 that severe testing is equivalent to maximising degree of confirmation, this analysis interestingly shows that extremely confirming strategies are not very sensible either. This is so for the same reasons as why severe testing should be performed moderately. Failing to find the very strong confirmation searched for can put one further back than obtaining a low value confirmation.

At first sight, there is a contradiction between our finding that testers in the rule discovery task feel unable to falsify on the one hand, and that they apparently control the chance of falsifications by means of severe testing, on the other. Selecting severe rather than weak tests, however, does not necessarily reflect a motivation to falsify. The purpose of such a strategy might be to give a strong proof for one's theory; the implication that the probability of falsifications increases by doing this, might subjectively not be felt so. This might explain why participants in the rule discovery task feel unable to force falsifications, even if they could by performing severe tests. Severe testing is not primarily associated with falsification, but with strong confirmations.

Alternative philosophical models for the rule discovery task. I round off the discussion of Wason's rule discovery task by considering an investigation in which unusual philosophies of science are applied to it. Tukey (1986) analysed the behaviour of testers in the rule discovery task using philosophical norms other than the norm of falsification. He wondered whether or not the test strategies that testers follow could be described and/or legitimated by means of philosophies of science other than falsificationism. Tukey applied Mill's guidelines for experiment and the theories of Lakatos, Kuhn, and Bayes, in order to predict testing behaviour.

According to Mill's methodology, one should search, among all the different possible characteristics of the rule, for the one feature that is a necessary and sufficient condition of the rule. In order to discover this, one should examine each feature to see whether or not it also occurs in the rule. If all possible features can be eliminated except one, the correct rule has been found. Testers should, therefore, construct various hypotheses (possible features), test these one by one or in groups, and present their hypothesis when all except one have been eliminated. Lakatos's theory describes scientific research as a series of tests in a research programme. Analogous to this, Tukey suggested regarding tests carried out by testers as being a research programme with a hard core. Thus, test behaviour should not be viewed at the level of individual hypothesis testing, but at the higher level of a series of test programmes with this type of hard core.

A Kuhnian framework, according to Tukey, makes the following aspects of testing behaviour important. The refutation of a hypothesis in Wason's task may be compared to a Kuhnian "crisis situation" in science. Wason (1960) found that testers sometimes tested the same hypothesis again after a refutation. While Wason regarded this as non-normative, it can be described as "rational" crisis behaviour in Kuhn's terms. It is possible, in fact, to experiment in a crisis without having a clear hypothesis. On the other hand, testers can perform "normal research" in another stage. When analysing Wason's task in the light of Bayes' theory, Tukey predicts that the tester will proffer the solution when his or her certainty concerning the rule has reached a personal criterion. In contrast to Mill's theory, not all alternative hypotheses need to be eliminated. It is sufficient when one hypothesis overtakes the other, in terms of subjective certainty, once it has become adequate to be presented as the solution.

Tukey replicated Wason's task twice. In the first experiment, he asked testers at the beginning of the task to generate as many hypotheses as possible which might "explain" the triad presented, after which the testers performed the tests. Tukey regarded the strategies of the testers as conforming to Mill's methodology when the hypothesis that they submitted as the rule was the only one that remained when all other alternatives had been refuted, and when this hypothesis itself was not refuted by the results of the tests performed. Tukey found that 80% of the testers did in fact behave in this way, and concluded that Mill's methodology provides a reasonable description of participants' behaviour. In a second experiment, Tukey found that people, just as in Kuhnian crisis situations, occasionally perform tests without having an explicit hypothesis. In addition, people appear to consider "confirming" and "exploring" to be useful test strategies in the rule discovery context. Finally, according to Tukey, testers appeared to behave in a way predicted by Lakatos, inasmuch as they do indeed test entire research programmes as well as individual hypotheses.

Tukey's attempt to use standards of testing other than those corresponding to Popper's philosophy is interesting but it also illustrates the difficulty, already noted in the case of Popper's philosophy, of squeezing these philosophies into the strait-jacket of the rule discovery task. The "experimental" translation of global concepts belonging to the philosophy of science, such as "research programme" or "normal science", is always debatable. Tukey's translation of Mill's philosophy is probably the most interesting. This strategy turned out to be very well used. Also, it resembles the strategy that has been found in several variants of previous rule discovery experiments: Participants successively eliminate alternatives, starting off by rejecting only a few, continue to search for some confirming evidence, and then proceed to eliminate more until some strong confidence in the surviving hypothesis is obtained.

SUMMARY AND CONCLUSIONS

The rule discovery task has revealed a number of mechanisms underlying testing behaviour which are alternatively labelled as rational or defective. The theoretical tools to describe these mechanisms are borrowed from theories of testing, notably Popper's falsificationism. In the present chapter, I have discussed this experimental work by focusing on the translation of the theories of testing into the experimental task, and by evaluating the resulting psychological claims.

"Confirmation bias" is a major finding. Testers tend to be oriented towards confirmation of their hypotheses at the cost of potential falsifying information. This was Wason's very first conclusion. Later, much of this claim was stripped down, precisely because of the operational definition of "confirming testing behaviour". Rather, the strategy displayed by participants in this task is more one of "positive testing behaviour", which does not necessarily reflect a wish to confirm or avoid counter-evidence to the hypotheses. Positive testing is a useful heuristic in many cases. It logically leads to the discovery of false positives alone and therefore to the narrowing down of one's hypotheses, at the cost of false negatives and the broadening of the hypotheses. However, this is not a biased strategy *per se*. It may be precisely what we want in some everyday testing situations. People seem able to switch to negative testing when the situation makes it a more efficient strategy. I also investigated the usefulness of actual refutations in the testing process. Clearly, being a successful falsifier does not guarantee discovery of the truth. The following overall picture emerges from the literature on the rule discovery task. In the first stages of the testing process, when the tester is still uncertain about her guess, confirming evidence for the initial hypotheses is pursued. After this has attained some plausibility, and the tester is reasonably confident regarding its truth, its limits are explored by searching for falsifications. Severe tests are performed. Finally, when the tester is very confident, the interest in falsifications disappears again. This strategy had some rationality in the rule discovery context. Participants who were very uncertain did not benefit much from falsifications.

A widely recognised rational strategy is the testing of alternatives. This is verified in the rule discovery task. Severity of test, the Popperian principle, did not turn out to be the best way to discover the truth. A test is more severe as more alternatives are excluded. Interestingly, testers acting severely got more falsifications but were less aided in finding the true rule. This result illustrates the asymmetry principle. Severe tests are relatively good "confirmers" but bad falsifiers. This also illustrates the unified confirmation-and-falsification principle. Severe testing leads to high degree of confirmation for the same reason that severe tests do. Severe testing (and therefore degree of confirmation) may be best when used moderately and at the right moment rather than maximised.

It is remarkable that the rule discovery task, covering the whole testing process, from test choice to result interpretation, has received less attention than the selection task in which the observed behaviour is limited to test selection. For example, in contrast to the selection task, very few attempts have been made to devise realistic versions of the rule discovery task. An exception is Gorman's (1995) application of the positive and negative testing strategies, arising from the Wason paradigm, to a natural discovery setting: Bell's work on inventing the telephone. In line with a number of philosophies of science, he suggests that negative and positive testing, as well as confirmation and falsification, should be considered in relation to the different levels at which a scientist works: At the molecular, experimental level a researcher might perform *negative* tests, in order to *confirm* some theory at the molar level of his research programme.

An explanation for the higher popularity of the selection task compared to the rule discovery task, might be precisely the extreme simplicity and shortness of the task as it is presented to the participants, in contrast to the huge complexity of interpretation of its results by scientists. This imbalance is much more dramatic than in the rule discovery task, which is also simple to administer and complex to interpret, but is, in addition, a more complete simulation of complex real-life testing situations. It is precisely this contrast, and the richness of the predictions the experiments generate, that make Wason's work such a brilliant contribution to the field. As we shall see in the next chapter, there are striking parallels in the discussions the two problems generate. 'How valuable is Popper's falsification standard?' is also a question in the selection task. Is a response that is irrational at first sight, interpretable as being rational after all, according to some other theory of testing or some pragmatic standard?

CHAPTER FOUR

Wason's selection task

INTRODUCTION

Wason (1966) first reported his selection experiments in a section of an article dealing with inductive reasoning and concept formation, in the Piagetian tradition. The book in which this article appeared (Foss, 1966) gives a rather popular survey of the new developments in psychology at that time. Wason wished to illustrate a cognitive limitation, the "verification bias", which is analogous to "confirmation bias". The standard of reasoning was proposition logic, and the particular question was whether or not testers spontaneously applied the *modus tollens* reasoning pattern, which makes use of falsification.

The original task has stimulated an enormous number of replications and an extensive research literature. This task has become the basis of an intense debate concerning not just conditional reasoning, but reasoning in general. It has become the vehicle of claims about such various fields as hypothesis testing, deductive reasoning, inductive reasoning, pragmatic reasoning, deontic reasoning, and common rationality. In the present chapter, I shall trace the main theoretical developments that have been generated or supported by experiments using this task. First, the original theories are summarised. Next, content variations and the definition of falsifying behaviour are briefly discussed. Most attention will be given to the recent statistical inference approach to the selection task because of its merits for understanding hypothesis-testing behaviour. Finally, I integrate a number of explanations of the selection task under the overarching principle of relevance. Some conclusions are put together about the significance of the selection task for understanding hypothesis testing. For some excellent reviews of the selection task, see, among others, Evans (1989, 1991, 1993) and Sperber, Cara, and Girotto (1995).

ORIGINAL EXPLANATIONS OF SELECTION TASK PERFORMANCE

Wason's (1966) participants were presented with a series of cards. On the front of the card was a letter and on the back a number. Testers were requested to indicate which cards should be turned over in order to show that the researcher is lying when he utters the following statement: "If a card has a *vowel* on one side, then it has an *even* number on the other side". The standard version of the task later evolved to become the following: The testers are presented with four cards, one of which displays a vowel, one a consonant, one an even number, and one an odd number. Subsequently, the testers are instructed to specify which cards should be turned over in order to determine whether the above-mentioned hypothesis is true or false. The testers could indicate as many cards as they deemed necessary.

Symbols may be assigned to each card: The vowel is P, the consonant not-P, the odd number Q, the odd number not-Q. The hypothesis can then be translated into an implication of the form: "if P, then Q". The tests are cards (P, Q, not-P, not-Q); turning over each has two possible outcomes (see Figure 4.1; see also Evans, 1989).

The "correct" answer, according to Wason, is the selection of a vowel (P) and an odd number (not-Q). The truth table constructed from the implication "if P then Q" has four alternatives: P, Q; P, not-Q; not-P, Q; not-P, not-Q. From a strict logical point of view, all alternatives verify the implication except "P, not-Q". Since "P, not-Q" falsifies the hypothesis unconditionally and since the other results can only provide equivocal verification, the participants will be expected to select the falsifying cards, in line with Popper's ideas. The cards P and not-Q should be turned over (Evans, 1982; Wason, 1966, 1968). However, only 10% of the participants opt for this solution, and this is a result that recurs in almost all replications of the standard task. Of the remaining 90%, most choose to turn only the card with the vowel (i.e., P alone), or the card with the vowel and the card with the even number (P and Q) in order to test the hypothesis. Wason (1966, 1968) concluded that participants do not recognise the value of falsification.

What then differentiates the selection task from the rule discovery task, which was also about falsification behaviour? First, there are differences relating to the very task the participant is faced with. This task is reduced to selecting a card (a test) only. In the rule discovery task, the tester performs

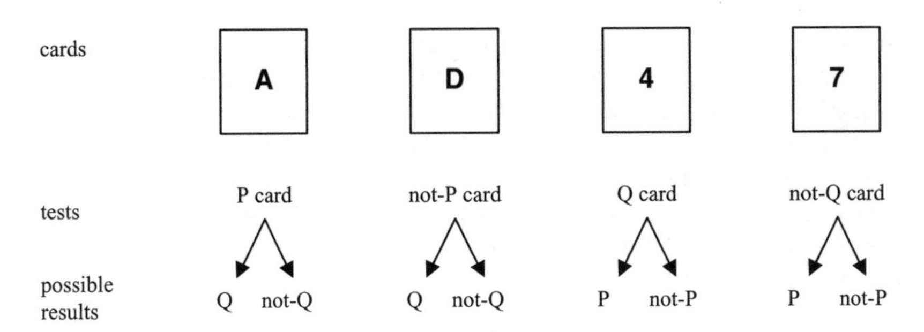

Figure 4.1. The logical structure of Wason's selection task (after Evans, 1989).

the whole cycle of a testing procedure: choosing a test, observing its result, and generating new hypotheses. We will see below that the very reduced character of the selection task leaves room for a great diversity of explanations, modelling, and theorising about the participant's intentions and estimations. Second, there is an interesting comparison to be made between the normative behaviour in these tasks. Consider the following argument. The hypotheses generated by the participants in the rule discovery task can be reformulated as a conditional statement: "If I propose any triple a–b–c satisfying my hypothesis H, then the experimenter will say 'yes'". The normative way of testing this hypothesis was proposing a *negative* triple: a triple a'–b'–c' not satisfying hypothesis H.

In the selection task, the choice is between a P card, a Q card, a not-P card, or a not-Q card, or a combination of those. The normative answer is the P and not-Q cards. However, in the context of this analogy, the not-P choice would be normative in the rule discovery task. Indeed, the triples satisfying H are P, and the yes-answer is Q. In the materials of the rule discovery task, there is no possibility of choosing Q or not-Q as a test of H, but there is a choice between P and not-P. The normative answer in the rule discovery task makes clear that the tester should be aware that her hypothesis can bear on more cases than the P cases alone, and that she should check this by means of not-P tests (possibly broadening her hypothesis). This could be seen as checking false negatives. Conversely, in the selection task context, in which the statement should be interpreted as a conditional, this not-P choice is not informative. Indeed, the statement cannot be falsified by means of not-P cases, because the relation mentioned in the statement is supposed not to tell anything about cases other than P cases. Therefore, not-P cases must be assumed to be irrelevant to the normative task performer testing the statement. Instead, she should check for "false positives": P cases that are not-Q.

This comparison illustrates that the selection task, at least originally, is a special case of hypothesis testing, because of the special logical constraints put on the hypothesis to be evaluated, and furthermore is underpinned by a normative model that occasionally conflicts with other standards that are adequate in other hypothesis-testing situations. This property of the selection task will be argued in the following sections to be a problem in generalising it to realistic hypothesis-testing situations. However, the recent analyses and findings of selection task research have loosened these logical constraints on which it is based and, by doing this, have contributed to the explanation of general hypothesis-testing behaviour.

What is at stake in the selection task?

Let us review some arguments about the mental processes supposed to be involved in the selection task. We may ask what exactly is observed. The task clearly involves a very restricted part of a usual hypothesis-testing situation. The participant does not actually perform the test, nor is she requested to evaluate the hypothesis after an actual test result is observed. Since the behaviour itself is so restricted, the response can be interpreted in numerous ways. The fascinating aspect of this task is that it is experienced by the participants as being very easy, whereas it is extremely difficult to interpret for psychologists. One of the most frequent descriptions of the task is that it tells something about human "conditional reasoning" (Wason, 1968) or "deductive reasoning" performance (Evans, 1982; Johnson-Laird & Byrne, 1991). In other words, the task is supposed to trigger the participant's (implicit) knowledge of formal proposition logic, which involves reasoning patterns such as *modus ponens* and *modus tollens*. In this approach, the participant's choice is described in terms of these reasoning schemes.

It has been argued, however, that the task is not actually a conditional reasoning task (Sperber et al., 1995). Proposition logical reasoning typically starts off with some premises and consists of deriving a valid conclusion according to the reasoning rules of proposition logic. In the selection task, however, this is not the participant's task. She is not asked to derive what can be concluded from the conditional and some other premise, but is asked to verify whether or not the rule is true by selecting cards that afford information about which premise is true. It is presumed that participants use some conditional reasoning in making this selection, and that there is a valid answer according to logic, but it is not certain that participants actually become involved in such reasoning when making a selection. The selection task could be placed in between pure reasoning tasks and hypothesis-testing tasks. Indeed, it differs from deductive reasoning tasks in that it has an "empirical" component. Not all the information needed is available for evaluating the conclusion, it has to be checked against reality. On the other hand, the task deviates from regular hypothesis-testing situations because of the special character of the hypothesis. It is a proposition with a connective, of which the truth status is assumed to be derivable from the rules of logic.

Another question, related to the foregoing one, is how appropriate is proposition logic for modelling and analysing the behaviour of naive participants performing this task? Proposition logic is an artificial system designed to analyse typical mathematical and formal statements. For example, it is the basis of a number of computer programming languages. But is it suitable for analysing natural language and natural reasoning? The connectives in proposition logic do not necessarily match the corresponding words in natural language. This match is particularly debatable in the case of material implication, which is the connective used in the statement. The meaning of "if . . . then" in logic is formally *defined* by its truth table, and does not always mean "if . . . then" in natural language. For example, one of the characteristics of material implication is that if the consequent is true, the conditional is always true whatever the truth value of the antecedent. Thus, "if P then Q" is *always* true if Q is true, in logic. Another one is that when both antecedent and consequent are false, the statement is true. Is this how we understand "if . . . then" in real life? If so, the following sentences should be immediately felt as being true by a naive reader:

(1) If a bird can fly, then Bill Clinton has been president of the United States.
(2) If a giraffe is a bird, then Bill Clinton has been president of the United States.
(3) If a giraffe is a bird, then Jacques Chirac is the president of the United States.

In the natural context, some understandable relation is generally assumed between the antecedent and the consequent. Without such a relation, the truth of the sentence is experienced as not able to be judged, and has little communication value (as the reader has undoubtedly felt when reading sentences 1 to 3). An additional problem for the application of proposition logic in the analysis of natural-language situations is that this logic is binary: It has two truth values. A sentence can be false or true in logic and nothing else. Therefore it cannot capture the intuition people have in natural language that some statements are neither true nor false, but "not able to be judged" or "perhaps true" or "probably irrelevant".

The problem with describing Wason's task as a conditional-reasoning task is the assumption that "if . . . then" is (or should be) read as a material implication. Few people will be familiar with the "grammar" of proposition logic. And even if they are, it is, of course, questionable whether they will interpret "if . . . then" in logical or in natural-reasoning terms when faced with this task. Even if the conditional is quite abstract in Wason's task, it is plausible that participants will activate the interpretation of "if . . . then" that is usual in natural communication. This supposition is supported by Hoch and Tschirgi (1985), who provided their participants with additional cues about the antecedent-consequent relation that were redundant to the *logical* antecedent-consequent relation in the statement. This version of the selection task, which encouraged an interpretation matching the logically correct interpretation of the "if . . . then" statement, indeed facilitated correct selections.

One may also put the problem as follows: "If . . . then" is "truth functional" in proposition logic. That is, if we only know whether the antecedent and the consequent are true, the truth value of the conditional is directly derivable. As can be intuitively seen in sentences 1 and 2, this is not the natural language case. In real-life reasoning, the truth of the sentence also depends on factors other than the truth of the two parts alone (see also Evans, Newstead, & Byrne, 1993). In sum, the very question of what kind of process is involved and consequently which theory should be used to explain the response has not yet been definitely settled. But the sheer amount of experimentation with this task has generated numerous new questions and findings regarding hypothesis testing, as will be shown in the following sections. In addition, the fact that this task is so often used, in spite of the objections one can have against it, has the clear advantage that the results of all studies are readily comparable, and each manipulation effect can be evaluated separately.

Logical rules

The first explanation Wason (1966) proposed for the preference of P cards and not-Q cards was provided in terms of proposition logic. People have a "defective truth table" for the material implication reasoning scheme. According to this defective truth table, the cases in which the antecedent is false (not-P) are considered irrelevant for the evaluation of the statement. Also, the participant correctly knows that the "P, Q" combination implies that the statement is true and the "P, not-Q" combination implies that the statement is false. The preference for checking the P card and the Q card is a preference for the "true" evaluation and avoidance of falsification, according to Wason. For the participant, the P and Q cards can provide verifying information, and not-Q can give either irrelevant or falsifying information (with not-P or P). Wason (1966) called this the verification bias. Thus, the response can be explained by the joint influence of a defective truth table and the preference for verifying rather than falsifying a statement. This preference was explained, in turn, by a lack of insight into the value of a falsification. This verification bias is obviously a problematic theory, because it does not explain why participants choose the P card, which can possibly falsify the hypothesis (Evans, 1989; Gadenne, 1982).

In a series of experiments conducted in the late 1960s and early 1970s, efforts were made to encourage the participants to choose the falsifying option by means of elaborate instructions. These were the so-called "therapy experiments" (Wason, 1968, 1969; Wason & Johnson-Laird, 1970), in which it became evident that the participants were very persistent in their refusal to turn the not-Q card in order to test the hypothesis. If the choice is incorrect (P only, or P and Q), the researcher invites the participant to indicate which consequence the possible results on the back of the cards would have for the hypothesis. The participants then evaluate the possible results, and are allowed to reconsider their selection. It appears that the participants are capable of providing a correct *evaluation* of the results when the test has been carried out: They state that not-Q would falsify the hypothesis. Nevertheless, this instruction does not result in the participants *selecting* the not-Q card as an instrument for testing the hypothesis. Wason (1977) calls this "self-contradiction".

Johnson-Laird and Wason (1970) formulated the first model of behaviour in the selection task, based on the human capability for using logical principles. Three levels of insight of the testers are recognised: no insight, partial insight, and complete insight. In the first case, the testers have no insight into the value of falsification and only select the P card or the P and Q cards, depending on their interpretation of the conditional. In the second case, the tester recognises that he or she must also choose possibly falsifying instances. The P, Q, and not-Q cards are then chosen. In the third case, the tester finds only the falsifying instances relevant. He or she turns over the P and not-Q cards. These levels of insight were deduced from the protocols of testers performing the task. In sum, when the task is considered as a test of human reasoning abilities, the results show that these are very poor. However, as O'Brien (1993, 1995) also shows, the task is a quite complex example of this logic in which many reasoning steps are required to be made ahead, and possible results compared. This can be seen in the logical decomposition of the task displayed in Figure 4.1.

Heuristics

However, do participants really make a logical "pre-posterior" analysis in which each card is considered in terms of the logical value of its possible result for the evaluation of the hypothesis? Evans proposed an alternative perspective. Evans (1972, 1982, 1983, 1989, 1998) explained the choice of the testers in terms of a "matching bias". Testers simply turned over the cards with those letters stated in the hypothesis to be tested. Wason (1968) and Wason and Johnson-Laird (1970) had already indicated this phenomenon.

This model is known as the "dual-process" model because it assumes two sorts of thinking processes: a logical and a non-logical process. The test selection is based on a non-logical matching heuristic; the subsequent justification of this selection is a rational, analytic process.

Evans and Lynch (1973) presented their participants with statements having a negated consequent: "if P then not-Q". This became: "If there is a vowel on one side of the card, there is not an even number on the other". When asked to select cards to test this statement, most participants preferred the Q card (even number), which was the *falsifying* card in this case. Hence, the "insight" and "verification bias" explanation was disproved. Indeed, if participants had made a pre-posterior analysis directed at finding the cards that would provide them with a verification, they would have selected the verifying case: the not-Q card instead of the Q card. On the basis of the experiments with negated statements, Evans (1984, 1989, 1993) argued that participants display a matching bias. This bias means that the selection is made on the basis of a simple judgement of the relevance of the cards. This relevance judgement is determined by the mere mention of "P" and "Q" in the rule to be tested, in the standard version of the task. However, in the negated version of Evans and Lynch (1973), the mentioned consequent is "not-Q". The fact that the Q card is assigned relevance rather than the not-Q card is due, according to Evans (1984, 1989), to the linguistic principle that a negated value in the consequent directs attention to that (positive) value. The new, basic perspective introduced here is that participants do not get involved at all in any logical pre-posterior analysis of the cards (the tests), but that they make their selection on the basis of a heuristic involving relevance judgements.

Wason and Evans (1975) used the negative version (if P then not-Q) of the task and, in addition, they examined the participants' protocols in order to study the reasons for selection. They also found that the negative version led to the normative card selection (the P and Q cards, in this case). But the testers' protocols seemed to be subsequent rationalisations of their selections rather than reflections of logical insight. They found that testers given the negative version of the task and who presented the logically correct solution by coincidence, due to the matching bias, gave "logical falsification" as the reason for their choice more often than testers given the affirmative version of the experiment. The general idea of Evans' matching bias and two-stage model has been supported in several replications of the selection task (Beattie & Baron, 1988; Jackson & Griggs, 1990; Platt & Griggs, 1993; Ormerod, Manktelow, & Jones, 1993).

Sperber et al. (1995) proposed an explanation of the selection task by means of the theory of relevance, which is a more global theory of communication and information processing (Sperber & Wilson, 1986) than Evans' relevance concept. However, this relevance theory is quite similar to

Evans' explanation of relevance (Evans, 1994), although Sperber et al. deny that the relevance judgement of a card is a heuristic. According to them, judgements of relevance are the product of implicit inference processes. The judgement that any particular information is relevant depends on two considerations. First, the greater the "cognitive effect" resulting from processing the information, the greater its relevance. Second, the greater the "cognitive effort" required for processing the information, the smaller its relevance. In short, a piece of information becomes more relevant as it is easier to process and more informative. Applied to the selection task, this means that facilitation of the not-Q selection is obtained when the not-Q case is made "salient and easily available". Sperber et al. argue that this can be generally realised when "one causes the participants to interpret the rule as a denial of the occurrence of P and not-Q". In other words, a context should be created in which knowing whether or not there are "P-and-not-Q" cases is informative, and in which "P-and-not-Q" cases are easy to represent. Sperber et al. argue that these conditions are satisfied in the negated consequent version of the task of Evans and Lynch (1973). Indeed, the hypothesis "if there is a vowel on one side, there is no even number on the other side" makes the even-number case salient and easily available. The interesting aspect of Sperber et al.'s relevance theory is that it explains not only the effect of the negated rule but also provides general conditions under which facilitation is obtained. For example, it also explains facilitation by content, which will be discussed in the next section.

The two-stage model based on (heuristic) relevance judgements and matching bias followed by an analytic reasoning process, presents parallels with hypothesis-testing behaviour analyses in the rule discovery task. Interestingly, Poletiek (1996) found that participants instructed to conduct a confirming or a falsifying strategy interpret this instruction as making a match (positive test) or a mismatch (negative test), respectively, between their hypothesis and the test item they propose, in order to satisfy the instruction. Thus, the testers do not make a pre-posterior analysis about the possible test results and their values right from the beginning of the test choice process, but they choose the relevant test items simply on the basis of what the instruction prescribes and the content of the hypothesis. Also, Poletiek found that, in a second stage, after the test item had been selected, an analytical reflection could be activated concerning the actual falsifying or confirming power of the chosen test item.

The third perspective (after the original explanations in terms of verification bias and the matching approach) on the results with the standard version of the selection task is simply that people possess a mental logic and apply it in the task (Braine, 1978; Johnson-Laird, 1983; Rips, 1983). This perspective is frequently disputed because it implies that people can apply these rules independently of the content of the reasoning problem. However, different inferences are made with different semantics. There is also much experimental evidence that inferences depend on context, as will be shown in the next section (see Evans, 1989; Evans & Over, 1996a; see also Manktelow, 1999, for a review).

Mental models

In addition to the defective truth tables, the heuristic perspective, and the straight use of mental logical rules, yet another position is taken by Johnson-Laird and Byrne (1991) in their theoretical account of the selection task. In their view, a tester faced with the selection task makes a "mental model" of the statement proposed by the experimenter. The construction of mental models is a general principle that plays a role in several forms of reasoning behaviour, according to Johnson-Laird (1983). It consists of "constructing an internal model of the state of affairs that the premises describe". These models are built on the basis of general knowledge. They are dependent on the context and content of the premises. The mental models generated can be both "explicit" and "implicit" models. Consider the selection task with the statement: "If there is an A on one side then there is a 2 on the other side of the card". It yields an explicit model representing the situation that for every card with an A there is a 2. However, since this statement is also compatible with the situation that there is no A but there is a 2, this possibility is generated as an alternative possibility that is not thought out explicitly but is represented as some fuzzy possible, implicit alternative. The participant considers only those cards that are represented in the explicit models. Up to this point, the reasoner is not involved in any reasoning. Building models is a matter of using general knowledge. In the final stage, those cards are selected for which the hidden value could have a bearing on the truth or falsity of the rule represented in the mental model. This process can be seen as a context-independent reasoning process in which rules are applied. Indeed, it demands some pre-posterior reasoning. The possible outcomes of different selections must be imagined. In order to produce an "insightful performance" it is thus necessary to construct a model in which negative instances are explicitly represented. Making implicit mental models explicit is called "fleshing out a model". Johnson-Laird and Byrne (1991) explain facilitation in the selection task by proposing conditions under which particular models are built and fleshed out. These conditions are, for example, changing the form of the rule, its context, its content, using negated terms in the statement, and altering the instructions as to ask explicitly for people to look for "violations" of the rule.

These conditions under which the negative-instance models are fleshed out call to mind those proposed by previous studies with selection experiments in order to explain facilitation of the selection of not-Q cards. For example, rules with a negated antecedent or consequent tend to elicit attention to the corresponding positive item (Evans, 1989; Evans & Over, 1996a; Sperber et al., 1995). Platt and Griggs (1993) showed that three manipulations of the standard selection task can dramatically increase a tester's performance: First, instructing participants to search for violations instead of merely to find out whether the rule is true or false; second, explicating the material implication rule; and third, asking participants to give reasons for their selections and non-selections. Platt and Griggs argue that these manipulations make it more likely that participants flesh out implicit alternative models in which the not-Q card plays a significant role. Ormerod et al. (1993) manipulated the form of the rule. One of the rules examined had the form "P only if Q" as the logical equivalent to the "if P then Q" rule. This enhanced performance. Braine (1978) had already proposed that such an "only if" statement emphasises the contrapositive case: "if not-Q then not-P". In the mental model theory, this is rephrased in the assumption that participants construct two explicit mental models from the beginning: one with the P in the antecedent and one with not-Q in the consequent. No implicit models have to be fleshed out. Since both models are explicitly available, the not-Q card is also selected (Johnson-Laird & Byrne, 1991).

In sum, the attractiveness of the mental model theory is its generality. It can explain a large range of phenomena in deductive reasoning. It has the advantage of explaining several abstract versions of the task as well as content variations, which will be discussed in the next section. Many theories on hypothesis testing merely focus on one of these. On the other hand, the assumption that people generate mental models with the properties which Johnson-Laird and Byrne (1991) describe is difficult to verify and falsify. For example, it is difficult to establish whether the selection of Q when testing a hypothesis with a negated not-Q consequent is a linguistic "bias" which focuses our attention on the actual word that is negated, or whether it occurs because the not-Q expression makes us "build a mental model" of Q. The mental-model explanation of the response in the selection task has more striking similarities than differences compared to other explanations, such as Evans' heuristic/analytic approach and Sperber et al.'s (1995) relevance explanation. Although Johnson-Laird and Byrne (1991) interpret Sperber and Wilson's (1986) relevance theory of reasoning as a theory of "mental logic", the application of the latter to the selection task in Sperber et al.'s (1995) paper displays an interesting overlap with the mental model explanation. Sperber et al. propose that the search for possibly falsifying information in the selection task is facilitated when this information is made "relevant to the participant". And a piece of information becomes relevant when it is "easy to represent" and as it has more "cognitive effect" (Sperber et al., 1995). These situations correspond largely to the situations which in mental-model theory predicts facilitation of the falsifying card selection. A model is fleshed out when it is triggered by a certain context or content, but then it is also easy to represent, and informative. In this manner, the mental models theory and the relevance theory can both explain a range of facilitation effects in the selection task with concepts that have been theoretically worked out differently, but which converge in the global concept that the falsifying case is actively looked for when it is activated in the mind, due to some manipulation of the task.

In the next section, one such type of manipulation is discussed. It has generated a comprehensive programme on pragmatic hypothesis testing as opposed to logical hypothesis testing.

CONTENT VARIATIONS AND FALSIFYING STRATEGIES

Since Wason and Shapiro (1971) first performed a study using a four-card task which involved an everyday situation instead of letters and numbers, an enormous number of "realistic content" replications have been performed. Although the volume of the programme would justify full deliberation, this discussion will mainly focus on its relevance for hypothesis testing, and how it resulted in new contributions from a statistical-inference perspective. Manktelow (1981, 1999) and Evans (1983, 1989) provide very good overviews of the programme.

Wason and Shapiro (1971) found that significantly more testers selected the P and not-Q cards when the problem was presented as an everyday one rather than as the classical version of the problem. The cards showed "Manchester" (P), "Leeds" (not-P), "train" (Q), and "car" (not-Q). The hypothesis to be tested was: "Every time I go to Manchester, I travel by train". The participants tested this hypothesis by turning over the "Manchester" and "car" cards. These are the cards that can produce a falsification of the hypothesis. Since this manner of task presentation appears to lead to a rational testing strategy, there was talk in the research literature of a thematic effect: Realistic task contents facilitate rational testing behaviour. Johnson-Laird, Legrenzi, and Legrenzi (1972) found that even more people performed better with the following task contents. The hypothesis to be tested was: "If the envelope is sealed, it has a 50-lire postage stamp". The participants had to imagine that they were Post Office staff who had to check this rule. The cards presented were in the form of an envelope: a sealed envelope (P), an open envelope (not-P), an envelope bearing a 50-lire stamp (Q), and an envelope bearing a 40-lire stamp (not-Q). Of the 24 testers, 22 indicated that they wished to turn over the P and not-Q cards. Notice that, given the instructions, this version of the task was not really a truth-evaluation task, but rather a deontic rule-violation-detection task. We come back to these versions later. Subsequent to these spectacular results, a series of experiments followed in the 1970s and early 1980s, in which the thematic effect was also obtained (D'Andrade in Rumelhart, 1980; Griggs & Cox, 1982; Pollard, 1981; Van Duyne, 1974, 1976). Van Duyne (1974) used a hypothesis involving a student's course of study and the city in which he is studying. Van Duyne (1976) created an interesting variation in which he allowed testers to generate a hypothesis themselves and to test it according to the task pattern. He discovered that the facilitating effect was greater when the tester said in advance that the hypothesis was *sometimes* true than in the case of a hypothesis which the tester was very *certain* was true. Thus, Van Duyne found a link between the subjective certainty about the hypothesis and the test strategy. This will be discussed again when I consider the statistical models. Griggs and Cox (1982) used the hypothesis: "If a person drinks beer he must be older than 19". This hypothesis had the format of a rule to be obeyed. This, also deontic, variation also led to a strong facilitation of the P and not-Q choice.

However, the effect of task contents came under fire, in the light of theoretical deliberations and critical replications. Manktelow and Evans (1979) performed a series of four-card experiments with realistic task contents, but discovered no effect. Moreover, they replicated Wason and Shapiro's experiment exactly and found no effect. Griggs and Cox (1982) also found no facilitating effect either with the contents involving transport and cities, or with the postage stamp experiment. Reich and Ruth (1982) and Yachanin and Tweney (1982) did not find the thematic effect in their replications. The criticism of successful content variations concerned the details of the experiments on the one hand, but was fundamental on the other. The first argument was that, in many instructions, the tester was explicitly encouraged to falsify. For example, the instruction given by Johnson-Laird et al. was: "Select those envelopes that you definitely need to turn over to find out whether or not they violate the rule". However, the findings concerning the effect of the instruction are not uniform. They have been refuted in a number of experiments (see Evans, 1989), and supported in others (Platt & Griggs, 1993). Interestingly, the very first selection experiment also had a variant on the violation instruction. After all, Wason (1966) instructed his testers to discover whether the researcher was "lying" with the hypothesis, but it did not work as facilitation.

The second argument is more fundamental: Various sceptics maintain, with regard to the content effect, that the content element in the presentation of the selection task leads to the task no longer appealing to logical reasoning, but more to memory of the situations described in the task. Memory then produces only the cues required to solve the task directly. The stage of logical reasoning is bypassed. This explanation, called "memory cueing", has been elaborated by Griggs and Cox (1982). Experiments with familiar contents with various groups of testers who had, or did not have, the relevant situation in their memories, confirmed this. Golding (1981) performed an experiment of this type in which she presented the postal problem. She discovered the facilitating effect to be present in the case of testers above 45 years old who were acquainted with this scrapped regulation, but to be absent in those who had not known the regulation.

The experiment carried out by D'Andrade (in Rumelhart, 1980) was an opportunity to place the memory-cueing perspective in proportion. Testers were placed in the imaginary situation of being the manager of a Sears store. Their task was to check receipts. Receipts in excess of 30 dollars must have the signature of the department manager on the back. The tester sees four receipts: one for 15 dollars (the not-P card), one for 45 dollars (the P card), one with (the Q card), and one without (the not-Q card) a signature. D'Andrade noted 85% correct answers when the tester was requested to check whether the rule had been applied. D'Andrade's result contradicts the explanation that testers only need to employ their memories in order to discover the correct answer, since they probably had never been manager at Sears and were not acquainted with the rule. At most, it is probably necessary to recognise the structure of the problem as the structure of a known problem. D'Andrade's experiment thus leads to a sort of middle position: The increase of the number of correct answers cannot be explained solely by reference to known realistic situations, while realistic contents do lead to better reasoning. Thus, there must be a process of reasoning involving a certain amount of abstraction while, at the same time, it is one which is activated by a concrete context.

This position can be found in Cheng and Holyoak's theory (1985) concerning pragmatic reasoning patterns. A reasoning pattern is more abstract than specific knowledge resident in memory, but more specific and concrete than rules of logical reasoning. This type of pattern is activated by the correspondences, in terms of content and structure, which it has with the concrete problem. According to Cheng and Holyoak, the fundamental pattern of the selection tasks with facilitating contents is the "permission pattern". The characteristic feature of this pattern is that it contains a criterion and an action and four different "moral" ways in which these can be linked to one another. One of these is, for instance: If the criterion is not fulfilled, the action may not occur. The postal problem and the drinking problem can be described according to this pattern. It appears that this pattern generates the same selection by the testers as the selection which is regarded as being logically correct in Wason's four-card problem. Cheng and Holyoak (1985) support their theory with experiments in which they illustrate that facilitation in the selection task only takes place if the contents conform to the features of the pattern.

Cosmides (1989) explained the facilitating effect by stating that successful variations in contents should be regarded as invoking social contracts.

She proposes an evolutionary approach to reasoning. The fundamental theory postulates that people are able to survive due to their capacity to interact socially. This works in conjunction with the ability to uncover those who violate the social contract, the "cheaters". According to Cosmides, the conditional hypotheses comply with the following basic rule: If you profit (in a social contract), you must pay the costs. The transgressions are the "P, not-Q" cases. Cosmides gained evidence for her point of view in a number of experiments. She showed that only those rules which can be translated into social contracts lead to facilitation. This evolutionary explanation of the facilitation effect has been criticised in a number of studies, e.g., Cheng and Holyoak (1989); Manktelow and Over (1990). These authors find that not all facilitating scenarios can be regarded as social contracts, as Cosmides defined them. The Sears problem is an opposite example. Manktelow and Over (1990) propose that the crucial feature causing the facilitation in previous variations of the task is the deontic character of the rule. The antecedent or consequent of the rule contains an action and the proposition generally contains a modal term, such as "must". Manktelow and Over conducted an experiment in which testers had to "test" deontic propositions. The propositions were "conditional obligations". These are deontic propositions in which a certain course of action is required under a certain condition, for example, "if you clean up blood, you must wear rubber gloves". They predicted that when high costs were connected to violation of the rule, the tester would choose the not-Q card more often. Their prediction was verified. According to the authors, the deontic character of the rule and the importance of transgression caused the facilitation, but without invoking social contracts. Interestingly, at this point, utilities are introduced in the explanation of test selection. The crucial factors that influence test behaviour are the costs and benefits of the rule "being true". The experimental manipulation of the perspective of the tester has underscored the importance of utilities in the explanation of realistic test situations. Indeed, different cards are turned over depending on whether one is cued into the perspective of the party who can be cheated or the party who deceives another party (Gigerenzer & Hug, 1992). In fact, the tester's perspective had already played a role in the experiments of Johnson-Laird et al. (1972) and D'Andrade (in Rumelhart, 1980), in which participants were cued to believe that they were staff members checking whether or not an employee had correctly applied a rule. Clearly, inducing testers to imagine a deontic testing situation in which they are actors elicits subjective utilities that, in turn, influence their test choice.

According to Manktelow and Over, the researcher using the selection task should pay careful attention to the subjective importance that the tester assigns to the hypothesis that is to be tested (1990, p. 163).

> We have already indicated that some new research on the selection task needs to be concerned with people's utilities and subjective probabilities.

Manktelow and Over introduced the metaphor of the "action to be taken" for the "truth evaluation of a hypothesis" (Chapter 2). This parallels the application of this metaphor in the statistical theory of hypothesis testing, especially the decision theory. But, along with the introduction of decision theory in the selection task research, the problems of this metaphor, which were discussed in Chapter 2, arise in the present context too. The major problem is whether or not people are still involved in hypothesis testing and truth evaluation when the context of the "testing" situation is deontic, and the actors are oriented towards actions. Manktelow and Over (1995) argue that deontic reasoning has its own logical form and content and that the psychology of deontic reasoning can best be studied by means of the theoretical concept of subjective utilities (Manktelow & Over, 1992). For instance, they criticise the mental model explanation of Johnson-Laird and Byrne (1991; see also Johnson-Laird & Byrne, 1992) for falling short of explaining pragmatic versions of the selection task, precisely because the mental model account does not include any reference to utilities. In other words, their position is that the deontic task triggers thinking that is different from that involved in the merely logical reasoning task. The separation between deontic versions and abstract versions of the selection task is now widely accepted by most authors. Testing for the truth or falsity of a hypothesis involves reasoning different from checking whether or not some (social) standard has been violated. However, the utility metaphor and statistical theory have gained influence in the selection task independently of the deontic programme. The major recent contributions (Friedrich, 1993; Kirby, 1994; Oaksford & Chater, 1994b) to the explanation of the selection task and other hypothesis testing research all take the perspective of hypothesis testing as statistical inference, abandoning the logical reasoning standard.

The statistical-inference perspective on the selection task has introduced two major assumptions, the first of which is uncertainty. In modelling the response of test selection in a task even as abstract as the selection task, a two-truth-values logic is no longer assumed, but probabilities. For example, "if P then Q" may be tested differently, depending on the participant's estimation of the probability of the antecedent and the consequent separately being true (Oaksford & Chater, 1994b). The other assumption is utility, in line with the statistical decision theory of hypothesis testing (Friedrich, 1993; Kirby, 1994). Notice that the introduction of utilities in this research implies that test results have values, and that the assessment of these values can determine testing behaviour in a particular testing context. Thus, a confirmation is not merely a confirmation but can be strong or weak, convincing or not, and practically relevant or not.

The very idea of utilities was already present in some theories of the abstract selection task, which are not explicitly based on statistical models. For example, Sperber et al. (1995) argue that the choice of the not-Q card can be facilitated by making the P and not-Q cases "easy to represent", "contentious", or "undesirable" (Sperber et al., 1995). There is but a short theoretical distance between making an outcome undesirable and increasing the tester's "utility" in detecting that observation. Thus the idea of utility has already been implicitly introduced in the explanation of testing behaviour in the selection task. But in the statistical inference approach, these utilities are explicit and mostly quantified. In the following section, the statistical analyses are discussed, along with their capacity to account for abstract and concrete versions of the selection task and human hypothesis testing in general.

STATISTICAL INFERENCE AND THE SELECTION TASK: A GOOD MARRIAGE?

The application of utilities to the selection task is quite intuitive in the case of deontic versions. Indeed, one can easily imagine that it may be more or less important to detect some kind of violation of a rule, depending, for example, on the practical costs of violations. However, can abstract versions of the task be modelled as statistical-inference tasks as well? Two major papers (Kirby, 1994; Oaksford & Chater, 1994b) have recently introduced this perspective and have offered explanations of many important findings with the selection task. Moreover, they gave birth to a new theoretical paradigm that generated many new predictions and models, not only about this task (see for example Klauer, 1999), but hypothesis testing in general. Therefore, the Kirby and Oaksford and Chater model will be discussed in detail, and the work following this line of research will be addressed more globally. I shall first focus on Kirby's analysis and Over and Evans' (1994) discussion of his paper (see later). Subsequently, Oaksford and Chater's probability approach and its critics will be discussed. These discussions will serve as a point of departure for an overall evaluation of the statistical-inference perspective.

Kirby's probability and utility model

Kirby (1994) proposes that testers in Wason's task select cards to turn over on the basis of the utility they attribute to some outcomes and the probabilities of finding such outcomes on the back of a card. This is an alternative to the logical approach (1994, p. 2):

> Subjects' choices may not be generated directly by a faulty proposition logic, by a "natural" logic, or by context-dependent procedures.

In this framework, the selection is based on the expected utilities: The utility of each outcome of a card is multiplied by its probability of appearing and the products are added together. This provides the expected utility of selecting the card. Kirby proposes an inequality that describes the criterion for selecting a particular card.

$$\frac{p(\text{inconsistent outcome present}|c)}{p(\text{inconsistent outcome absent}|c)} >$$

$$\frac{u(\text{inconsistent outcome absent; c not selected}) - u(\text{inconsistent outcome absent; c selected})}{u(\text{inconsistent outcome present; c selected}) - u(\text{inconsistent outcome present; c not selected})} \quad (4.1)$$

This criterion basically maintains that if the utility (u) of an inconsistent outcome on the back of a card (c) weighted by the probability (p) of such an outcome is higher than the utility of a consistent outcome weighted by its probability, then the card will be selected. Importantly, in this formula, the utility of *selecting* an *inconsistent* outcome (falsification) is positive and the utility of *not selecting* a *consistent* outcome is positive as well. Also, locating an inconsistent outcome is called a "hit". In other words, Kirby assumes that participants try to locate falsifications. In the framework he proposes, this means that detecting them has a positive utility. Thus, in Kirby's proposal, the utilities are defined beforehand in accordance with the logical normative analysis of the task; participants should pursue falsifications. And it is assumed they do.

However, we "lack information about the magnitudes of the utilities that any participant assigns to the possible outcomes" (Kirby, 1994, p. 5). Therefore, Kirby only manipulates the probabilities, keeping the utilities constant. Kirby's participants can calculate the probability that a card will reveal a falsification. This card will be more preferred as this probability increases, as Kirby hypothesised. In Kirby's stimulus materials, the probability of a falsification with a not-Q card increases, as the set of P cards increases as compared to the not-P set. Thus, he predicts that more not-Q cards will be selected as the size of the P set increases. Over and Evans (1994) paraphrase this prediction in a more intuitive way. For the statement, "if it is a raven, then it is black", the participant will search for inconsistencies among the ravens rather than among the not-black things (the raven set is "small"). Studies similar to the ones Kirby performed had already previously been conducted (Pollard & Evans, 1981), but Kirby proposed the first comprehensive analysis in statistical terms.

In a typical experiment, Kirby told his participants that a computer had printed out 100 cards with an integer on one side of the card and a + or a – on the other side of the card. Depending on the condition, the participants had to test the following statements:

- If the card has a 1 on one side, then it has a + on the other side (small P-set condition).
- If the card has a number from 1 to 50 on one side, then it has a + on the other side (medium P-set condition).
- If the card has a number from 1 to 90 on one side, then it has a + on the other side (large P-set condition).

Subsequently, the tester was told that the computer makes one mistake in about every 10 cards. Participants were asked to select those cards that had to be turned over to determine whether or not the statement holds true. Kirby found that more participants selected the not-Q card as the P set increased, as predicted. He also found that fewer participants selected the P card as the P set increased. The latter finding was unpredicted. Kirby does not explain it, but Over and Evans provide an interesting interpretation, which will be discussed further on in this section. In the following, the focus is on the relation between the size of the P set and the not-Q selections. Subsequently, the explanation for the relation between P-set size and the selection of P cards will be analysed.

The P-set-size effect on not-Q card selections. Before dealing with Kirby's set-size phenomenon, I shall briefly reflect on the falsification assumption Kirby made. An important aspect of Kirby's model is that it assumes that participants try to find inconsistent cards. That is, he assumes that finding falsifications has a positive utility. Accordingly, the explanation he gives for participants failing to select not-Q cards in the original vowel–even numbers version is that participants neglect this card because they estimate the probability of finding a falsification on the back to be very low. However, trying to explain why participants are or are not actually interested in inconsistencies, being the most informative observations according to proposition logic, was originally the very purpose of the task. Thus, the signs of the utilities attributed by Kirby *a priori* to the different outcomes (consistent or inconsistent) are precisely the "*explanandum*" in the selection task. The explanation that Kirby provides explains falsifying card selections by participants who are assumed to pursue falsifications on the basis of their estimations of the probability of obtaining them. The very question of the actual utility of these outcomes is not dealt with (Kirby, 1994, p. 4).

> Discovering the relative utilities of consistent and inconsistent information awaits future research and is not discussed here.

Thus, the important new aspect of Kirby's analysis is about set-size effects, which relates to probabilities of the elements mentioned in the hypothesis. I will argue that the set-size effect Kirby shows is task specific, due to the particular selection task materials he used. However, in his line of thought, more general set-size influences in the selection task are plausible. These general set-size influences, which I will demonstrate, might also *explain* why testers prefer the falsifications. In other words, it might explain the utility of falsifying cards felt by testers.

The gist of Kirby's explanation for the P-set-size effect on the not-Q selections is that participants prefer not-Q cards as they have a higher probability of providing a falsification. In Kirby's experimental manipulations, the increase in the set size of P does indeed lead to an increase in this probability. Kirby does not fully demonstrate this relationship. (For a demonstration, see Appendix 2.) To provide this demonstration, and thus to calculate the probability, some additional information is required. It is necessary to know how the not-Q cards are distributed in the not-P set (see Appendix 2). Kirby gives no information about this. More generally, this information is almost never given in variations of the selection task. For example, in the standard version ("If there is a vowel on one side of the card, there is an even number on the other side"), nothing is said about the distribution of the uneven numbers over the consonants; or in the raven version ("If it is a raven, then it is black"), we do not know how many non-black things there actually are among the non-ravens.

Let us assume that the probability of a not-Q card in the not-P set is .50. The probability of detecting a falsification with the not-Q card indeed depends on the set size of Ps in relation to the number of not-Ps. As the P set gets larger, the probability that the not-Q card will issue from the P set increases, simply because the probability of it issuing from not-P decreases. For example, if there are 100 P cards and no not-P cards, all possible not-Q cards are necessarily from the P set. In Figure 4.2, the probability of getting a falsification with a not-Q card is displayed as a function of the P-set size.

The relationship between P-set size and the card preference proposed by Kirby is an interesting phenomenon in the selection task. However, the stimulus materials on the basis of which this effect was found deviate in some respect from the typical format of the materials used in the selection task. Notably, the set sizes in Kirby's version are known, as well as the "error rate". This implies that, in Kirby's materials, the P-set size also influences the probability of the statement being true. Indeed, the greater the P-set size, the smaller the probability that the computer has made no printing errors, in a particular output. And if no printing errors are made, the statement is true.

As opposed to Kirby's version, the statements to be tested in the typical case are generally utterances about infinite or extremely large sets; conditionals are often universal statements. Either the mentioned set or its complement is specified, although very large, and the other one is unspecified

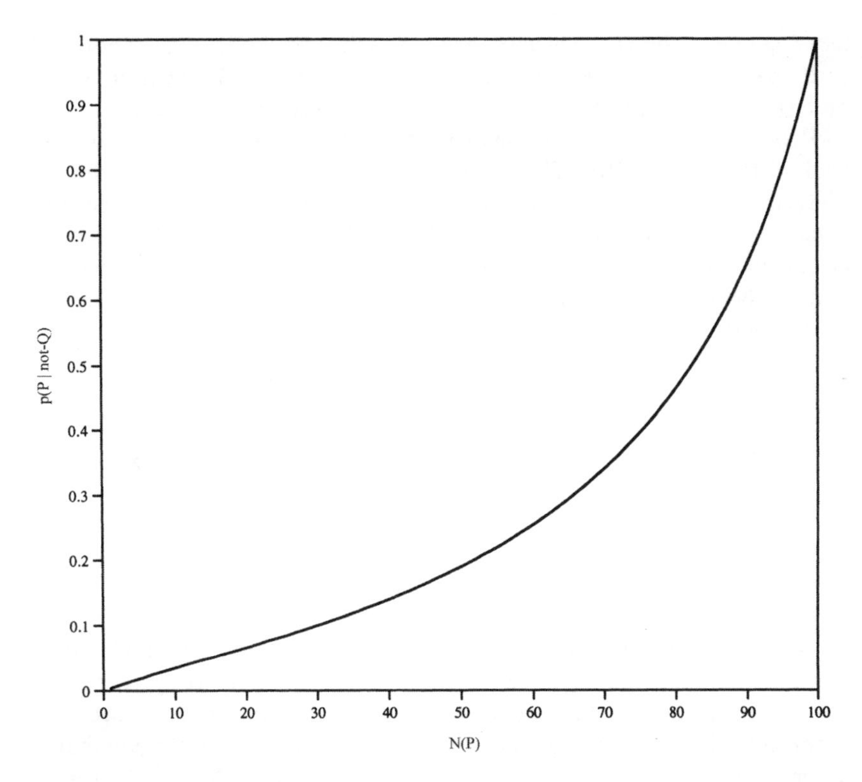

Figure 4.2. The probability of getting a falsification by means of turning the not-Q card as a function of the set size of P, in Kirby's (1994) materials.

and even larger and fuzzier. For example, the statement, "all ravens are black" is about ravens (a big and specified set), not-ravens (a much larger and unspecified set), black things (a very large set), and not-black things (a much larger and unspecified set). If we compare the size of the set of ravens to the size of black things, we feel that both sets are enormous but the set of ravens is the smallest. However, size differences in the complementary sets are very fuzzy: The set of not-ravens is so large that there is hardly any difference between it and the extremely large set of not-black things. Apart from the size of the P and Q sets, the Kirby experiments deviate from the general case in that, in the latter, no precise information is given concerning the prior probability of the statement.

Is it possible to make an analysis and to predict selections on the basis of set sizes for this more general case? Let us assume that conditional statements are generally about two sets that are specified but extremely large, with two complements that are infinitely large. Let us also assume, in line with Kirby and propositional logic, that testers attribute a positive utility to locating falsifications. They can do this by selecting P or not-Q or both. The card that has the highest probability of revealing a falsification will be preferred. The probability that P will be found on the back of not-Q ($p(P \mid \text{not-}Q)$) is compared to the probability that not-Q will be found on the back of P ($p(\text{not-}Q \mid P)$). (The probability that a result is found on the back of a card is equivalent to calculating the probability that a certain card will issue from a certain set, as shown in the problem above.) These values can be calculated by means of the absolute set sizes as follows. The probability of a falsification with the not-Q card is the ratio of not-Q cards in the P set to the total number of not-Q cards:

$$p(P \mid \text{n-}Q) = \frac{N(P \cap \text{not-}Q)}{N(\text{not-}Q)} \tag{4.2}$$

The probability of a falsification with the P card is the ratio of the number of not-Q cards in the P set to the total number of Ps:

$$p(\text{not-}Q \mid P) = \frac{N(P \cap \text{not-}Q)}{N(P)} \tag{4.3}$$

Both these ratios have the number of instances that are P and not-Q in the numerator. This number is constant, whichever test we perform, but unknown. The denominators contain the number of instances of not-Q and P, respectively. Those are the set sizes of P and not-Q. It can be seen that the tester should select the card from the smallest of both sets N(P) and N(not-Q) if he or she wants to maximise the probability of finding an inconsistency. Interestingly, in this calculation, the set-size effect is not due to the increase or decrease of the set size in relation to its *complement*. This is because the statement is "universal"; about very large sets. The only thing the tester is doing, according to this model, is comparing the set sizes of P and not-Q, and choosing the smallest. As we have seen above, an intuitive *ranking* of their sizes is generally possible, even if they have a huge indeterminate size.

Thus, it might be possible to predict and explain set-size effects in the selection task with universal statements about almost infinite sets, when a search for falsification is assumed. This prediction is consistent with Kirby's finding. Indeed, he found that, as the P set increases, the probability of having a falsification with not-Q also increases, and not-Q cards will be preferred. However, in the present model, the size *difference* between P and not-Q explains the relative preference of P and not-Q cards, not the *absolute* sizes. This perspective needs further experimentation, but there are at least some past findings consistent with this analysis. Consider the black raven example mentioned by Over and Evans (1994), "if it is a raven, then it is black". Comparing the set of ravens with the set of non-black things, the first one is necessarily smaller. Thus, the P is preferred by the falsifier. But the negated consequent version of it can also be explained in this line of thought: "if it is white, it is not a raven" (see Over & Evans, 1994). The set of white things (P) is bigger than the set of ravens (not-Q), so in this situation, the preferred card for the falsification seeker is not-Q: the set of ravens. In summary, the present analysis is a variation on Kirby's—it focuses on a more general case in the selection task. It explains, in a more general way, relative preferences for the P and not-Q card as possible falsifiers. Interestingly, this analysis is consistent with the findings in the rule discovery task (Klayman & Ha, 1987). Indeed, Klayman and Ha (1987) explained the positive testing strategy as a search for falsifications in the smallest set, because falsifications are most likely to be found there (see Chapter 3).

In the previous analysis of general set-size effects, the positive utility of falsification is still assumed rather than explained, however. I propose that this very analysis points to a possible psychological explanation of falsifying behaviour in the selection task. In order to explain the positive utility of falsifications, we reverse the relation between the set-size effect and card preference. Indeed, it might be that differences in set sizes of fits containing possible exceptions *induce* a higher utility of falsifications. In particular, the size and the amount of specificity of a set involved in the statement might influence the ease with which looking for inconsistencies come to the participant's mind. Consider again the statement, "if it is a raven, then it is black". It might trigger the salience of exceptions within the ravens by the mere fact that ravens are the smallest and most specific of all mentioned sets: the P, not-P, Q, and not-Q sets (see also Nickerson, 1996). In the example, "if it has a vowel on one side, then it has an even number on the other side", the vowels are the smallest set represented by the participant. This might make exceptions among the vowels relevant. In the negated conditionals, such as, for example, "if it is white, it is not a raven", the smallest and most specific set this statement covers is that about ravens. This might induce a search for exceptions within this set, resulting in a facilitation of turning the raven (not-Q) card. In "if it is black, then it is a raven", such a facilitation might not be expected, because the set size of black things does not differ much from the set size of not ravens.

This "reversed" explanation of the set-size and utilities effects converges with the explanations mentioned in the previous section, the gist of which was that some versions of the selection task elicit relevance considerations. Following this line of thought, the set-size differences and the ensuing specificity differences might be another way to evoke the relevance of inconsistencies. Oberauer, Wilhelm, and Rosas Diaz (1999) proposed a similar explanation for preferences in the selection task: category coherence. Indeed coherent categories, as opposed to diffuse or unspecified ones, are easier to represent mentally, they argue, and therefore more often selected. Hence, set sizes become one of the explanations of facilitation, fitting in with the general theory of relevance. In sum, this analysis explains the positive utilities of seeking falsifications. As we have seen, merely assuming this is unsatisfying because of the amount of counter-evidence in other contexts than Kirby's. The analysis needs further examination, however. It may be a new step in the twist Kirby gave to selection task research.

In the following section, I proceed with the second set-size effect Kirby discovered, but left unexplained. The interesting aspect of this effect is that Over and Evans (1994) gave an interpretation in terms of their concept of "epistemic utility".

The P-set-size effect on P-card selections. P cards were less preferred when the statement was about bigger P sets. Notice, that this finding is globally consistent with our "reversed set-size explanation", that testers seek information in the smallest set. However, the relation between P-set size and P-card preference was an unpredicted finding of Kirby's experiments. Over and Evans (1994) introduced the concept of "epistemic utility" to explain it. In Evans and Over (1996a), epistemic utility is a mainly psychological concept. People choose cards according to their pragmatic or epistemic goals. The evidence that is perceived as useful for this goal, and that is available to the participant, will be chosen. In the context of the analysis of Kirby's results, however, epistemic utility is defined more precisely in probabilistic terms. Over and Evans (1994) argue that the epistemic value of a piece of evidence is related to its capability to make the statement more probable. In this manner, they hypothesise that the decrease of P-card selections is due to the decrease of epistemic utility of this card as the set size of P increases. Epistemic utility is defined be Over and Evans is as follows (1994, p. 239):

> We assume that the epistemic utility of an outcome is higher in one condition than in another if it makes the conditional statement proportionately more probable in the one than in the other.

The conditions referred to are the different P-set-size conditions in Kirby's experiments. They argue intuitively that turning over a P card in a large P set and finding a Q will increase the probability of the statement, but not by much since there are still many possible falsifications behind the numerous other P cards of the set. As the set increases, the number of possible falsifications also increases; therefore the contribution of one Q observation on the back of a P card becomes less as the P set increases. Over and Evans' intuitive argument is based on a Bayesian analysis. They do not, however, provide a full analysis. I propose such an analysis of this concept and its relation to set sizes.

It follows from the definition above that the epistemic utility holds for hypothesis-confirming outcomes only. In contrast to the assumption Kirby makes with his utility analysis in which testers are supposed to locate inconsistencies, Over and Evans propose, in fact, that the utility of a card increases as the statement *gains* probability. Thus, participants are supposed to locate the cards whose outcomes will confirm the statement to the greatest possible extent. A falsifying outcome cannot "make the statement more probable". The definition proposed by Over and Evans (see citation above) can be formalised as the ratio of the probability of the statement (H) after having observed the outcome, to the prior probability of the statement. It reflects how useful an outcome is in making the plausibility of the hypothesis increase. This is:

$$\frac{p(H \mid x)}{p(H)} \tag{4.4}$$

Notice that this is the same ratio as Carnap's relevance ratio (formula (1.3)), the measure for the degree of confirmation that a particular observation could contribute to a hypothesis, and the amount of revision of belief in the Bayesian framework (Chapters 1 and 2).

Does this measure relate to the set size of P? Assume a selection task situation. What is the epistemic utility of turning over the P card and finding a Q? The probability of the statement being true can be calculated by means of Bayes' theorem

$$p(H \mid Q \setminus P) = \frac{p(Q \setminus P \mid H) \cdot p(H)}{p(Q \setminus P)} \tag{4.5}$$

in which Q\P means "observing a Q on the back of a P card". The epistemic utility can be derived directly:

$$\frac{p(H \mid Q \setminus P)}{p(H)} = \frac{p(Q \setminus P \mid H)}{p(Q \setminus P)} \tag{4.6}$$

In the right-hand part of the equation Popper's measure of severity of test (formula (1.4)) can be recognised. Thus, the epistemic utility of observing a Q on the hidden side of a P card is not only formally equivalent to its degree of confirmation, but also to its severity-of-test. The equivalence between the two was already argued in Chapter 1 and is nicely illustrated here in the epistemic-utility analysis of the selection task. How does this measure relate to set sizes in general in the selection task, and how in Kirby's experiment?

We know that the term in the numerator in the right-hand part of the equation is 1. Indeed, the probability of finding a Q on the back of P is 1, given that the statement is true. Now, we need to calculate the prior probability of a Q on the back of P, with no prior assumption regarding the truth status hypothesis. "No prior assumption" is the same as assuming that "the hypothesis is either true or false". The participant's estimation of this probability will be higher as the card is an element of a larger P set. Indeed, she estimates how likely it is that "either there are only Qs on the back of Ps (the hypothesis is true), or there is *at least one* not-Q on the back of the P cards" (it is not true). Now, imagine that the statement is about a set of two Ps only: The expectation of finding a falsification not-Q if it is there on the back of the P card displayed, is "higher" than when the statement is about a set of 1000 Ps. Thus, the chance of observing a confirmation Q on the back of a P card is higher in the larger set, when we make no prior assumption.

But this "expectation" about what is on the back of the P card also depends on how much we believe the hypothesis to be true rather than false, beforehand. If we take into account this belief in modelling the tester's expectation of the chance of a Q on the P card, the expectation may be modelled as follows:

$$p(H) = [p(Q \setminus P]^{N(P)} \tag{4.7}$$

in which N(P) is the number of elements in the P set. In the long run, we expect the same proportion of Q with P as we believe the statement to hold.[1] Thus, the probability of a Q with P is such that:

$$p(Q \setminus P) = \sqrt[N(P)]{p(H)} \tag{4.8}$$

Assuming the prior belief p(H) constant, that is independent of the set size of P (N(P)), we can fill in formula 4.6. The epistemic utility of a Q on the back of a P card, as a function of the set size of P (N(P)), is:

$$\frac{p(H \mid Q \setminus P)}{p(H)} = \frac{p(Q \setminus P \mid H)}{p(Q \setminus P)} = \frac{1}{\sqrt[N(P)]{p(H)}} \tag{4.9}$$

This function is displayed in Figure 4.3.

Formula 4.9 implies that the *epistemic utility* of a confirmation with the P card decreases very steeply as the set size of Ps increases. This is because the *probability* of finding such a confirmation increases with P-set size. As the P-set size increases, we get what we already expect. Now, these calculations have brought us back to our unified principle of test severity and relevance, and prepare the way for a discussion of the probability value model later in the book: The more likely a confirmation *a priori*, the smaller its value, in general, for confirming the hypothesis. Thus, in the general case, we might interpret the preference for finding confirmations by means of P cards as a

---

[1] This is a debatable combination of frequentistic and subjective interpretation of chance, but in the context of this situation it might not be unrealistic (see Chater & Oaksford, 1999 for a discussion of these two versions of the probability of a hypothesis).

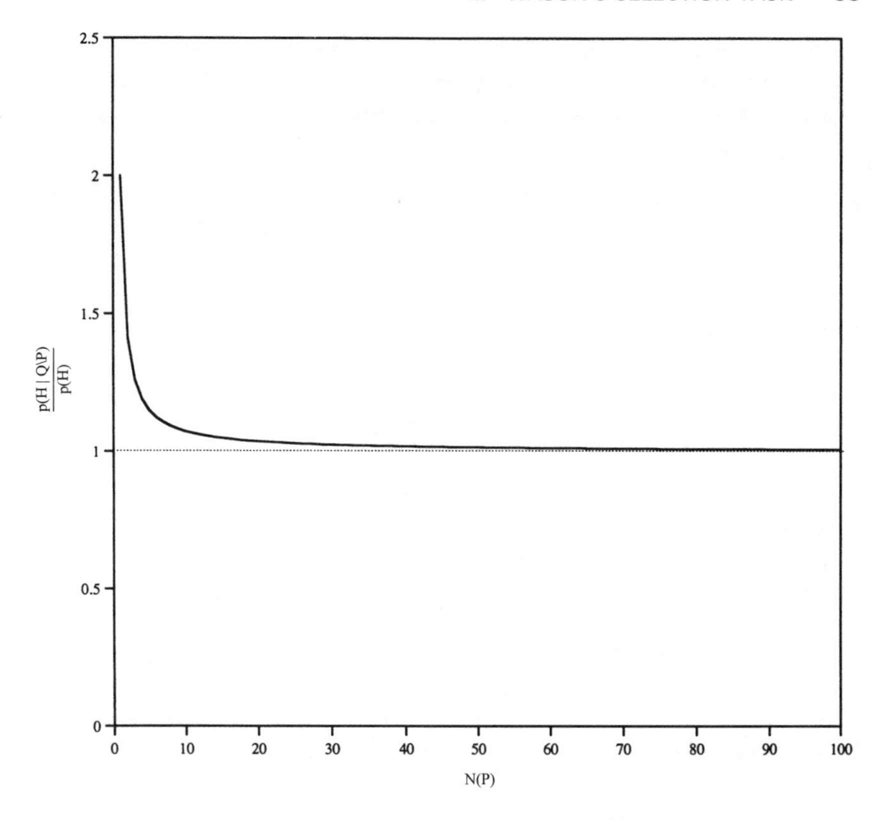

Figure 4.3. The epistemic utility of finding a Q card on the back of a P card, as a function of the set size of P.

set-size effect, because the P-set size can influence the epistemic utility, severity, and relevance of such a confirmation. In other words, searching confirmations among small sets might be interpreted as true severe testing behaviour in the selection task.

I now turn briefly from the general case back to the application of the concept of epistemic utility in the specific experiments of Kirby, as Over and Evans originally intended. A singularity in Kirby's materials is that the belief in the truth of the hypothesis, p(H), is directly related to the set size of P. This implies that one crucial assumption that we made in the foregoing analysis does not hold: The assumption that the belief of the tester in the statement is constant over different set sizes of P. Remarkably, this particularity (the dependency between prior belief and set size of P) eventually implies that the epistemic utility of finding a Q on the back of a P card is *constant* over the set size of P, and thus *independent* of set size. This can be demonstrated as follows. The prior probability of the statement (p(H)) in

Kirby's materials depends on the number of Ps. Indeed, as can easily be seen, the probability that the statement is true, i.e., that the computer has made no error in a large set of printouts, is very low, as Over and Evans also observe. This probability is:

$$p(H) = p(\text{no printing errors}) = \left(\frac{9}{10}\right)^{N(P)} \tag{4.10}$$

This probability decreases as N(P) increases. The probability of finding a Q value on the back of a P card in Kirby's experiment is; see also (4.8):

$$p(Q \setminus P) = \sqrt[N(P)]{\left(\frac{9}{10}\right)^{N(P)}} \tag{4.11}$$

which is 9/10. Thus, the epistemic utility is, according to formula (4.6), 10/9 *whatever* the P-set size is.

Consider the intuitive version of this argument. In a large set, we feel that finding a + on a P card does not contribute much to our belief in the hypothesis being true. Many falsifications are still possible, as Over and Evans explain. However, our belief that no printing errors are made was initially very small. In a small set, one with one P card, for example, finding a + is decisive. As shown above, *in relation* to what we already believed (our belief was already quite "strong" that no printing errors are made), it does not contribute much either. In all cases, observing Q on the back of P increases our belief with the same factor relative to our prior belief in the hypothesis. In sum, the formal version of the concept of epistemic utility (as we deduced from Over & Evans, 1994) does not adequately explain card selections, because it should predict no special preference for P cards when the P set is small, since under these conditions, such a choice does not enhance epistemic utility.

But how is it possible to explain the preference for P cards as the P-set size decreases, which Kirby found, if the formal epistemic utility explanation fails in this specific context? The preference for P cards in small P sets might be explainable by means of the utility of this card *to falsify* the statement rather than to confirm it (increase its probability). This is precisely Kirby's general idea. Remember that whatever the set size of P, the probability of finding an inconsistency when turning a P card is .10. This is the error rate that is directly given to the participant. If we display the probability of finding an inconsistency with a P card ((p(not-Q\P)), together with the probability of an inconsistency with a not-Q card (p(P\not-Q)) (which was shown in Figure 4.2), as a function of P-set size, the picture in Figure 4.4 results. It might be that participants consider both the not-Q card and the P card as possible falsifiers, and that they *compare* their relative power to

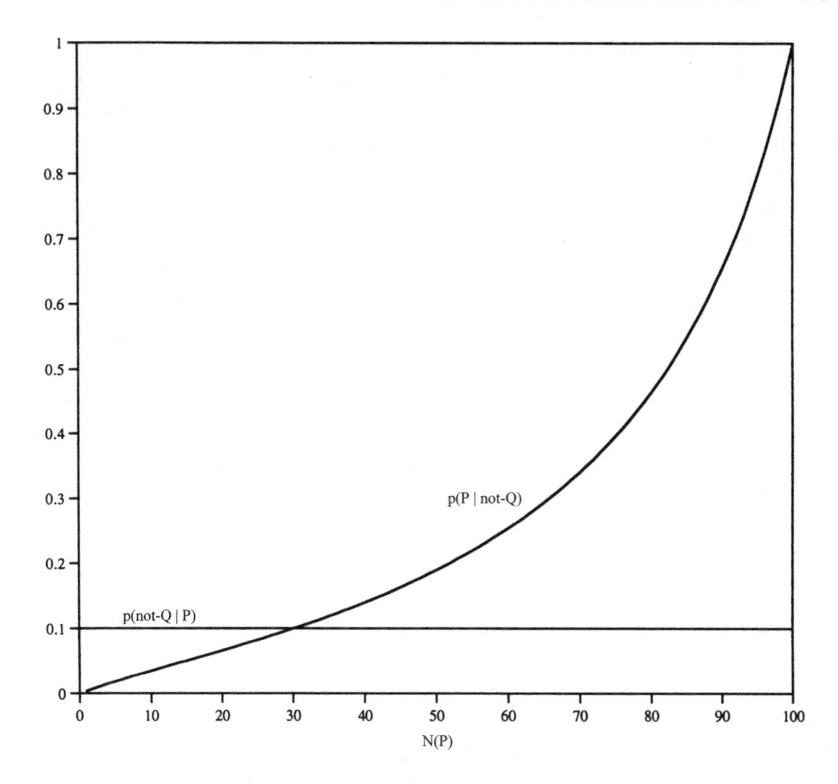

Figure 4.4. The probability of getting a falsification by means of turning the not-Q card, and the P card, as a function of the set size of P, in Kirby's (1994) materials.

detect a falsifying result. This is what I proposed in the previous section. If so, Kirby's finding is understandable. As can be seen, in Figure 4.4, the P card is a better choice than the not-Q card below the point at which the two lines intersect (under the assumption I accounted for previously, that the distribution of not-Q cards over not-Ps is .50). This is true for set sizes of P smaller than approximately 30. For larger set sizes, the probability of detecting an inconsistency is higher if a not-Q card is turned over rather than a P card. If the present analysis holds, Kirby's results should reveal an increase in preference for not-Q cards *together* with a decrease in preference for P cards. This is indeed what Kirby found (Kirby, 1994, Table 1, p. 8). Thus, participants' card preferences may depend on the comparison they make between several cards with regard to their ability to effect a falsification. The P and the not-Q cards are the only ones to be endowed with such falsifying power. After having pre-selected these two cards mentally, participants might compare them with regard to their probabilities of enabling the detection of inconsistencies, and make a final choice. Notice that, while this provides an explanation for the P-set-size effect Kirby found, again, it does not explain why participants search for falsification in the first place. At this point, the explanation proposed in the previous section might also underlie the present findings: Differences between the set sizes of P and not-Q might induce the search for exceptions in the smallest of those two.

Summarising, the Kirby experiments have generated the discovery of an interesting phenomenon in the selection task: set-size effects. The explanation Kirby offers, and subsequently also given by Over and Evans, is one in terms of utilities and probabilities. Which outcome has a positive utility depends on the theoretical assumptions: Kirby assumes that participants like to detect falsifications, Over and Evans assume that testers try to detect "useful" confirmations. However, the assumption that subjects try to locate one type of outcome is somewhat unsatisfying, because such an assumption is precisely what we would like to explain rather than simply assume in the selection task. It also leaves unresolved the issue concerning *when* one of the two assumptions should be made. After Kirby, the central question in the selection task remains: Do people look for falsifications or confirmations and what influence does set size have on these strategies?

One way of approaching this problem has been presented in the previous section. Perhaps the set-size differences within the statement direct one's attention to the smallest set, because this set is generally the most "specified" and therefore the most vividly represented. Oberauer et al. (1999) also point at both specificity and (small) set size as facilitators of card preference. An increase in selections of the P or not-Q card might be explainable by its relatively small set size, which induces the tendency to look for exceptions there. Another "statistical" manipulation that induces preference for not-Q cards is the suggestion that the statement has or can have exceptions; thus, "there may be Ps that are not-Q" (Love & Kessler, 1995; Manktelow, Sutherland, & Over, 1995; Pollard & Evans, 1981). The statistical inference explanation does not fully come to terms with this phenomenon. Indeed, should we assume that the presence of exceptions decreases the tester's belief in the overall statement (Chater & Oaksford, 1999; Oaksford, Chater, & Grainger, 1999)? A psychological explanation might be more adequate: It is quite straightforward that the enhanced availability of counter-examples, cases of P and not-Q, causes enhanced preference for these cards. These exceptions, once made explicit, may simply gain relevance because attention is directed at them, due to the instruction.

This and other examples of the assumptions in the statistical inference programme illustrate that the very application of this kind of statistical model to the selection task causes essential frictions. In "if . . . then" statements, falsifications and confirmations do not have the same status: For the statement to be true, all cases must satisfy it, whereas for it to be false, one exception is enough. Thus, these kinds of hypotheses cannot be confirmed or falsified "gradually". The same incompatibility problem emerges for the estimation of p(H) in the context of the selection task. In conditional reasoning, there is no way to represent more or less belief in the statement. In conclusion, it might be that the utilities and probabilities approach introduced here contributes most to explaining hypothesis-testing behaviour, and less to the selection task data it was intended to explain. I come back to this in the final section of this chapter.

The second main approach within the statistical-inference paradigm is the information theoretical approach. Oaksford and Chater (1994b) proposed that participants pursue "information gain" when selecting cards in the selection task. Their model predicts that most information is gained by selecting precisely the P and Q cards, when the P and Q sets are small. By this, they claim to provide a rational explanation of the selections that are most frequently found in abstract versions of the selection task and were previously labelled as irrational. Thus, as in Kirby (1994), set-size effects ultimately make a "rational" explanation of selections possible, when some statistical reasoning is assumed. In Oaksford and Chater's model this reasoning aims at "optimising information gain".

Oaksford and Chater's information gain model

In contrast to Kirby, who primarily focused on induced preference for the P and not-Q falsifying cards, Oaksford and Chater (1994b) aim at explaining the most preferred card selection: the P card and Q card. This selection, especially the Q card, is irrational according to proposition logic. However, Oaksford and Chater's statistical model predicts such a choice when a number of assumptions are satisfied. The general psychological assumption is that participants turn over those cards that optimally reduce uncertainty, or maximise information gain, which are equivalent in Oaksford and Chater's model. Information gain is defined as the "distance" between the "prior" uncertainty about a hypothesis and this uncertainty after some observation has been made. This value is calculated for all possible outcomes of an experiment (turning a card) and all hypotheses in the problem. The participant calculates this value for each card that she thinks about turning over. Subsequently, the cards with the greatest expected information gain are actually turned over.

The authors formalise the uncertainty of a hypothesis by means of the Shannon-Wiener measure of information. The uncertainty of a number of mutually exclusive and exhaustive hypotheses H_i (e.g., the statement is true (H_1) or false (H_2)) is defined as:

$$-\sum_{i} p(H_i) \cdot \log p(H_i) \tag{4.12}$$

The uncertainty measure for turning over a card and observing a datum x is the information gain. This measure is the "Kullback–Leibler distance" (Oaksford & Chater, 1996) between the prior (old) and the posterior (new) probability distributions of the hypotheses.

$$D(p^{new}, p^{old}) = \sum_i p(H_i \mid x) \cdot \log \left[\frac{p(H_i \mid x)}{p(H_i)}\right] \tag{4.13}$$

Now, the *expected* distance between the prior and the posterior distribution of the hypotheses can be calculated by taking into account each of the two possible outcomes (x_j) a card can provide.

$$ED = \sum_j p(x_j) \cdot \sum_i p(H_i \mid x_j) \cdot \log \left[\frac{p(H_i \mid x_j)}{p(H_i)}\right] \tag{4.14}$$

Interestingly, in the ratio epistemic utility and, again, the relevance ratio and severity can be recognised. Subsequently, Oaksford and Chater (1994b) define the two hypotheses that the participant considers in the selection task. These are: The statement is true, that is, P and Q are dependent on each other as the statement describes; or the statement is false because P and Q are completely independent of one another. The two hypotheses are assumed to be complementary and exhaustive. In the standard version, this would mean that the participant compares the two following models and only these. The dependence hypothesis (Oaksford & Chater call it Md) being "all vowels have an even number on the other side"; and the independence hypothesis (Mi) being "the vowels are equally distributed over the even and odd numbers as are the consonants". Another assumption is that the participant regards the number of Ps and the number of Qs in the absence of P, as being constant whatever the hypothesis. For the standard version, this would imply that the participant thinks that whatever the case (the statement is true or false), the number of vowels and the number of even number cards without a vowel stays the same. A third assumption Oaksford and Chater make is crucial for their rational explanation: They assume that the participant estimates the number of Ps, and the number of Qs in the absence of P, to be small (the "rarity assumption"). In the standard version, this means that both the vowels and the even numbers without a vowel are rare.

On the basis of these assumptions, Oaksford and Chater calculate the information gain which can be expected when turning over a P, Q, and not-Q card, the preferred cards in the selection task. Their model predicts that the information gain of a not-P card is always zero, because the log term is zero for all results (see formula 4.14). Not-Ps say nothing about the statement. Notice that this consequence is in line with proposition logic but not with the rule discovery task standard, which prescribes that one should try to broaden one's hypothesis by checking not-P cases. The three free parameters of their model are the probability of a P, the probability of a Q (in the absence of a P), and the prior probability of the hypothesis that the statement is true. The information gain with the P card is calculated by the authors to be the highest, and, surprisingly, the information gained by turning over the Q card was higher than that provided by turning over the not-Q card, in those regions where the probability of P and the probability of Q are small. This is obviously contrary to the logical standard. Oaksford and Chater demonstrate that this analysis can explain many results with the selection task as rational, abstract versions as well as realistic and deontic versions. I shall confine the discussion to Oaksford and Chater's model itself, some critical alternative models, and to its theoretical merits in explaining hypothesis testing in the selection task context.

Analysis of Oaksford and Chater's model, and some critical alternatives. The most criticised aspect of the information gain model is its psychological assumptions. Oaksford and Chater give no clear psychological justification of why their model and its assumptions correspond to what happens in the participant's mind when he or she is presented with the selection task (Evans & Over, 1996b). It can describe the data, especially the preference for Q cards in the various selection experiments, without giving a true psychological explanation, however. Why do people seek "information gain" and, more importantly, why in this particular way? This is especially problematic for the present model because of the number of assumptions it makes, and because of the formal character of the assumptions and the model. If these assumptions are not justified as psychological processes, then such a model might be suspected of giving an *ad hoc* description of the data. This, globally, is the criticism expressed by Laming (1996), who argues that the model assumes more than it can explain. Moreover, the model being mathematically coherent and describing the behaviour does not guarantee a "rational" explanation of this behaviour. In principle, all behaviour can be "covered" by such a model. This is why the focus is on the number and the plausibility of the assumptions, from a psychological point of view.

Before addressing the assumptions, let us compare the statistical perspectives on the selection task of Oaksford and Chater on the one hand and with those of Kirby on the other. A striking difference between the two is that Kirby assumes that participants try to locate falsifications. Subsequently, the model built on this assumption explains which card is preferred under different set-size conditions. Also, the epistemic utility model (Over & Evans, 1994) was interpreted as a model that assumes that participants try to locate the cards that can confirm the statement. It was argued that these statistical models (Kirby's and Over & Evans') explained set-size effects to some extent, but they did not explain why participants wish to falsify or to confirm. Now, Oaksford and Chater do not assume that participants pursue either falsifications or confirmations; they postulate that the outcome that maximises information gain has the highest utility. This means that the outcome that is expected to cause the most change in either of the two beliefs (the statement is false or it is true) will be chosen. This I will call the symmetric assumption of utility of a card, as opposed to Kirby's asymmetric assumption.

The symmetric assumption is attractive because we need not explain why the participant is biased towards one kind of logical result (confirmation or falsification). The disadvantage is that pursuing "information" is quite abstract for a psychological assumption and seems to require quite complicated calculations by the participant in the selection task. Indeed, the question, "which card should I turn to become best informed about all possible states of nature?" is more abstract in nature and possibly more complex to answer than the question, "which card should I turn to get a falsification?". The first question requires a mental calculation in which all possible outcomes of all possible cards (tests) for all possible hypotheses are taken into account (see formula 4.14). Even if we assume that this calculation is done implicitly rather than explicitly, it is quite demanding from a cognitive point of view. It might be more plausible that testers organise mentally the possible test outcomes in terms of confirmation and falsification of their belief. Another possible difficulty with a symmetric model is that no semantic influences are accounted for, which might direct our attention to either falsifications or confirmations. Both have equal status. Thus the statement, "all ravens are black" would elicit the same testing behaviour as "all swans are black". From the foregoing chapter we saw that there is some influence from the mental availability of exceptions on test selection, however.

An important assumption in the Oaksford and Chater model concerns the two hypotheses that participants consider. This is the statement itself on the one hand; that is, the P and Q mentioned in the statement are supposed to be related as the "if . . . then" statement describes. On the other hand, participants are assumed to consider the alternative that P and Q are entirely independent of one another. This means, for example, that for the statement, "if it is a raven then it is black", two hypotheses are considered. Either the statement is true, all ravens are black, or else the proportion of black ravens (not being 100%) is equal to the proportion of non-ravens that are black. But this alternative hypothesis is only a special case of all possible alternatives. The plausibility of this psychological representation of the two hypotheses depends on the content of the statement. In the example cited, a plausible alternative to the statement would be that there are a few exceptional kinds of ravens that are not black. But, in the Oaksford and Chater model this would imply that the tester also assumes that there are only a few exceptional things (not ravens) that are also not black. Thus, the assumed representation of the alternative hypothesis might be too specific in many versions of the selection task. A more general (and psychologically plausible) representation would be that there is "at least one exception to the statement", as was argued in the previous section on Kirby's set-size effects. Another assumption in the Oaksford and Chater model involves the restrictions put on the values of p(P) and p(Q | not-P) (Qs in the absence of P). These are constant regardless of the statement being true or false. The problem with this assumption also is that there is no information available to the participant about the relative frequencies of Ps and Qs (Laming, 1996). The statement only informs us about the relation between them. In sum, Oaksford and Chater's restrictions that are imposed on the parameters of the model might be reasonable in some cases, but not in others.

Apart from the psychological assumptions, the measures used by Oaksford and Chater have been thoroughly discussed in the reactions to their model. Next, I shall first discuss the use of information theory in this task, and subsequently two interesting alternative statistical-inference models, proposed by Laming (1996) and Klauer (1999). Evans and Over (1996b) and Laming (1996) have pointed towards several problems relating to Oaksford and Chater's application of the Shannon–Wiener concept of information to the selection task. For example, Laming argues that this measure is not a good metaphor for the selection task situation: It is intended to analyse situations in which a message is transmitted from a source to a receiver by means of a "channel" (Shannon, 1948 in Laming, 1996). The messages have a certain amount of information and the expected amount of information from a channel can be calculated on the basis of a number of observed messages. But there is no such information analysis possible in the selection task. However, in Chapter 2 I argued that information theory can be applied to hypothesis-testing situations. One confusing thing with the information metaphor as used by Oaksford and Chater might be that they apply the information function to *hypotheses* rather than *observations* (see formula (4.12)). Indeed, as shown in Chapter 2, the information value of an observation is inversely related to its probability of appearing, in analogy to the information value of a message being captured by a receiver. The analogy is that the more unexpected a message or an observation, the more informative it is (Chapter 2). However, the information value of a hypothesis (as Oaksford and Chater calculate, see formula 4.12) or an idea can hardly be represented psychologically. Indeed, we do not naturally feel that an idea that is probably untrue is more "informative" than one that is probably true.

As a reaction on Oaksford and Chater's information gain model, several alternative statistical models of the selection task have been proposed. I discuss two of them. Laming (1996) proposed another information theoretic approach. He also proposes to frame the selection task in terms of the information value of the possible outcomes. He recalculated the expected information values of each card taking the log-likelihood ratio (formula 2.5) as a measure of information, but "maximising the likelihood" in order to estimate the values in the log-likelihood ratio. This method is common in statistical-testing situations, especially in Neyman and Pearson's lemma (Chapter 2). This approach corresponds to considering an outcome and estimating its likelihood by taking its highest possible probability if the statement is true and if it is false. This is the maximum likelihood principle underlying the Neyman–Pearson lemma. For example, in the classical version "if there is a vowel, then there is an even number on the back", a participant considers the even number card (Q). If the outcome is a vowel, the likelihood $p(\text{vowel} \setminus \text{even} \mid H)$ is estimated to be 1. But the maximum-likelihood estimate of a vowel if the statement is false is also 1: $p((\text{vowel} \setminus \text{even}) \mid \text{not-}H)$. The same values are obtained for the alternative outcome: a consonant. Whatever the truth status of the statement, the maximum likelihood of this outcome is 1. Subsequently, the likelihood ratios for these outcomes can be calculated: They are 1, the log-likelihoods are 0 and the expected log-likelihood ratio is 0 as well. Thus, in this statistical analysis of Laming, the Q card has no information value whatsoever for the participant. It should not be selected.

Interestingly, the information value calculations of the cards with the maximum-likelihood model prescribe exactly the same selections as the classical proposition-logic standard. The P card and the not-Q cards should be selected. The Q card is not informative, in contrast to the Oaksford and Chater prediction. In the log-likelihood approach, the assumptions are not as restrictive as in Oaksford and Chater's model. There is no assumption made about parameters like the prior probability of the statement, or the rarity of Ps and Qs. However, in contrast to the Oaksford and Chater model, it does not explain the majority of the findings with the selection task. What is the psychological plausibility of the maximum-likelihood method? In principle, I think it is more intuitive and simple than the information-gain method. The mental procedure of thinking of the highest possible chance to see what we see under both hypotheses to be evaluated is quite reasonable. Imagine that we want to test whether all swans in the park are white and we see a number of white swans. We assess the highest probability in observing this group of white swans if they are indeed all white and if they are not all white. We infer that they are indeed all white if the former estimation reaches a certain criterion. The problem with Laming's analysis to the selection task is that apparently something goes wrong in this pre-posterior analysis with regard to the Q test, even if it is not complicated to perform. Participants do not reason that each observation made of the hidden side of this card has the same highest probability of appearing under all possible conditions. Thus, although the Laming model seems psychologically more plausible, it can only show that participants "apply" it wrongly, at least to some Q tests.

A second alternative to Oaksford and Chater's model is from Klauer (1999). Klauer actually proposes two ways of modelling the participant's behaviour in the selection task. In the first model, the participant is assumed to look for the test procedure that gives her most information about the truth of the statement after the smallest number of observations. In this model, the costs of testing determine the tester's card choice. This model is a "sequential" test model, because the tester considers what the optimal way is for choosing a number of subsequent tests. Klauer formalises this model by means of the "Kullback–Leibler information numbers" which is similar but not equal to the "Kullback–Leibler distance" used by Oaksford and Chater. The interesting difference is that Klauer's model assumes that, at each point in the test procedure, the Kullback–Leibler information number measure is maximised for the hypothesis that is currently most supported by the known evidence. Thus, it is a symmetrical measure in the sense that it aims at finding the truth about *whichever hypothesis* is true with lowest possible costs. But at each point in the procedure, evidence is searched for that has most power to confirm one of the two hypotheses, namely, the already most supported hypothesis. Klauer claims that this statistical model can explain the not-Q selection when the statement is "probably false" in the selection task (Fiedler & Hertel, 1994; Love & Kessler, 1995; Pollard & Evans, 1981, 1983). This can be intuitively understood as follows: Under this instruction, the currently most supported hypothesis becomes not-H. This is the hypothesis that she will try to confirm. Notice that, as Evans and Over (1996a) noted, this is basically a tendency to *confirm* the not-H hypothesis.

The second model from Klauer (1999) is non-sequential. The problem a participant in the selection task is faced with, is choosing the cards that will maximise his chances to correctly accept or correctly reject the statement (depending on what matters to him). It considers testing as reducing important decision errors. This model is asymmetrical because the tester can be concerned with either confirmations or falsification of his hypothesis. To investigate this model's empirical adequacy to describe actual card selections, Klauer estimated the parameters of this model on the basis of a large data set from selection task experiments. The fit between data and model was reasonable, but not completely satisfactory. There is, however, an interesting implication of this model test. Under conditions of rarity (as in the Oaksford and Chater model), and the tester's subjective belief that the statement is true rather than false, then the model fits best actual card selections when the tester "weights more heavily the error to falsely accept the rule when it is false than the error of falsely rejecting the rule when it is true" (Klauer, 1999). This means that belief in the statement is correlated with a tendency to avoid false confirmations. This conclusion is striking because we might expect that testers prefer to see a favoured statement confirmed in whatever way; i.e., even falsely. I will come back to this when discussing the probability value model.

At this point, we have a number of symmetric and asymmetric statistical models of the selection task: the Oaksford and Chater model, the Laming model, and Klauer's models. All models assume that participants want "information". Oaksford and Chater's information-gain model can explain a huge amount of selection-task findings, especially regarding the preference for Q cards. However, it makes assumptions that are complex, quite specific and psychologically difficult to justify. It is not intuitive (Laming, 1996). Klauer's models can also explain people's focus on not-Q cards when the statement is known to have exceptions. However, we argued in the context of Kirby's study the difficulty of estimating the p(H) when the tester knows there are "exceptions". Moreover, Klauer's model comes to these explanations by means of quite complex modelling as well, where a psychological relevance explanation might also provide a plausible explanation. Laming's statistical analysis based on the log-likelihood maximisation seems quite transparent. It is consistent with the Neyman–Pearson theory (Chapter 2). Importantly, it is theoretically consistent with the theory of conditional reasoning. Hence, it leads to the same predictions as the original proposition logic model that was the very theory for which the selection task was designed. However, it does not succeed in explaining the data. Thus, most statistical models are empirically powerful but difficult to grasp theoretically. Some are rather theoretical, but not empirically satisfying. Have we returned to the starting point again?

The conclusion inferred from the present debate about modelling the selection task statistically is the same proposed with regard to the Kirby discussion. The value of the statistical-inference models of Oaksford and Chater might be hidden rather than demonstrable with the selection task experiments. Indeed, the selection-task situation demands some quite specific assumptions due to the stimulus materials. For example, "if . . . then" statements say nothing about possible not-Ps that are Q, just by definition. Neither do they say anything about how the alternative state of affairs should be represented. This is why the assumptions that Oaksford and Chater nonetheless make are problematic. The statement just describes a relation quite formally in the language of proposition logic. But how representative are these "if . . . then" statements and the selection task for reasoning and hypothesis-testing situations in real life? The focus on the selection task has generated a huge amount of knowledge about information search, and reasoning. But it might no longer be productive as a source of experimental support for the presented theories. Maybe people do reason quite often in terms of information gain, but not in terms of "if . . . then" statements and cards to turn. The most interesting benefit of the statistical-inference approach might be not so much the explanation of the selection task data but rather its contribution to hypothesis-testing research in general.

RELEVANCE

In the previous analyses of the selection task, the explanations of card selections in varying versions of the selection task seem to converge in the principle of relevance. In the psychology of reasoning, relevance is defined in terms of the amount of cognitive effort needed to process the information, and the information's cognitive "effect" (Sperber et al., 1995) and in terms of its mental availability and its "epistemic utility" (Evans & Over, 1996a; see also Hardman, 1998). As a piece of evidence gains subjective relevance for the participant in Wason's task, the corresponding card will more frequently be chosen. However, relevance itself is not a final explanation of card selection. To have a full psychological explanation we need to know why some information is considered as relevant by the tester and other information is not. This is basically what the selection-task research has produced: The conditions under which some evidence becomes relevant and therefore searched for by the participants in this task. In the beginning of this chapter, a number of these conditions were discussed. In the second part of the chapter, a number of statistical effects of the task (p(P), p(Q), p(H)) were also explained in terms of relevance considerations.

The relevance explanation might thus be seen as a higher level explanation for the selection task, which might integrate a number of explanations that were independent and competing until now (see also Almor & Sloman, 1996). For example, deontic versions and versions in which the perspective or the interests of the tester regarding some information were shown to make this information relevant and therefore selected. Matching bias, which is a very powerful explanation for card selections in the selection task, might also fruitfully be included under the umbrella of the relevance explanation, rather than seen as an alternative to relevance. Indeed, matching the cards with the terms in the statement can be explained in terms of availability and minimising cognitive effort (Sperber et al., 1995). Also, the cases in which this matching bias does not operate, for example when the falsifying cases are formulated as explicit negations (Evans, 1998), they might be seen as situations in which the negated cases still gain relevance because their mental availability may be relatively high or the cognitive effort to process them as possible evaluators of the statement is reduced. Similarly, matching bias being suppressed in realistic situations might have to do with the fact that realistic items have a higher availability *per se* because they possibly elicit concrete mental images, ethical and aesthetical associations, personal experiences, etc., in addition to an abstract concept like a "vowel" or an even number.

Interestingly, the concept of relevance already emerged at the very beginning of this book when the philosophical theories of testing were presented. In particular, the relevance ratio expresses the impact of some evidence on the degree of confirmation of some theory. However, when this formal definition of relevance (which is equivalent to the formal definition of epistemic utility) is applied to selection task experiments, we run into problems, because the *formally* highly relevant observations are not necessarily chosen by selection task participants. The problem is that the relevance ratio values of the cards are fixed. For example, the probability of the statement being true when a P card with a not-Q on its back is observed is 0. This is because in the framework of the selection task, which considers the "if . . . then" statement as material implication, one falsification is enough to reject the statement definitely. This in turn is the very reason why the not-Q card has such a special status in this research programme. Thus defining psychological relevance by means of a formal measure like the relevance ratio leads to frictions in explaining card selections. There are, however, many good psychological relevance explanations of the selection task which explain precisely why the formally low relevant cards are preferred. However, we believe that the formal definition is a powerful tool for modelling information seeking behaviour in general hypothesis-testing situations, unconstrained by the rules of proposition logic, especially since this formal model is often intuitive in more natural cases. I shall come back to this.

The incompatibility between a formal definition and a psychological definition of relevance possibly explains why the definition of "epistemic utility" proposed by Evans and Over (1996a) and Over and Evans (1994) to explain selection task does not yet provide a full explanation. On the one hand it is a formal model (Evans & Over, 1996a; Over & Evans, 1994). This is meant as the normative ("impersonal") definition. On the other hand the authors distinguish a psychological "epistemic value" which is "what people do actually place epistemic value on" (Evans & Over, 1996a, p. 40). This psychological definition of epistemic utility converges with their definition of relevance. At this point, the explaining power of the concept of epistemic utility might suffer from its broadness.

In the second part of the present chapter, recent statistical analyses of set-size effects, as well as effects of the prior probability of the statement, were presented. It can be argued that these effects can also be interpreted as causing relevance effects. For example, it was hypothesised that the facilitating influence of (relatively) small set sizes for selecting the corresponding card might be a relevance effect. Indeed, relatively small sets are generally also more specified sets, which are more easily represented mentally than big sets. The rarity assumption (Oaksford & Chater, 1994b) and Kirby's (1994) set-size effects are consistent with this hypothesis. Other recent analyses go in the same direction. For example, Oberauer et al. (1999), following Murphy and Medin (1985), propose that not set size *per se* but category coherence affects selections because coherent categories are more easy to represent in a mental model. Evans and Over (1996a) propose that concretely described sets get more attention. Negatively described sets are generally very large (i.e., the set of non-ravens) which also affects the perceived relevance of checking this set when new information is searched (Evans & Over, 1996a). In other words, the crucial variable influencing card preference might be relevance by means of set size. Green, Over, and Pyne's (1997) experiments also support this. They showed that when the size of the sets of P and Q are manipulated independently of some other relevance cue, set size *per se* did not affect card selection. Oberauer et al. (1999) reported an experiment by Green and Over, who used a task that resembles a natural hypothesis-testing situation more than does the selection task. They proposed causal rules to be tested. Under these conditions, P and Q cards were preferred when they were infrequent. In sum, we believe that the set-size effects found in the selection task are interpretable as psychological relevance effects that affect general hypothesis-testing behaviour.

The influence of prior belief in the hypothesis, similarly, might be explained in terms of relevance. There are conflicting results about the influence of one's belief in a hypothesis on test choice. A number of studies, like Kirby's, manipulate p(H) by suggesting that the rule has exceptions. This makes participants select these exceptions more often, because their attention is directed to them by the instruction. However, when the actual subjective degree of belief in a hypothesis is manipulated, no effect on card selection is observed (Green & Over, 1997; Oaksford, Chater, & Grainger, 1999). A salient example illustrating that the effect of p(H) on test selection is mediated by relevance effects is the finding by Sperber et al. (1995) that a context in which the statement is uttered as absolutely true (very high p(H)) and explicitly claimed to have not one single exception, precisely directs the attention to these exceptions and makes them relevant. In sum, relevance was argued to be a powerful integrative explanation of the selection task (Evans & Over, 1996a; Sperber et al., 1995). This explanation not only nicely accounts for effects produced by the classical variations, like negated (Evans, 1998; Oaksford & Stenning, 1992) and deontic versions (Manktelow & Over, 1995) of the task, it also best covers the probabilistic effects that emerged in the statistical approach of the task. The statistical approach to the selection task takes into account uncertainties, prior beliefs, set sizes, and other realistic properties of hypothesis testing to explain what kind of information participants focus on to evaluate their hypothesis. By doing this, however, the statistical approach occasionally comes into conflict with the original task it aimed to explain. As I argued in the previous section, the marriage between statistics and proposition logic is not without problems.

However, where the marriage runs into problems, best service is lent to hypothesis-testing behaviour. Consider two comparisons between conditional reasoning and everyday hypothesis testing that illustrate this. First, Nickerson (1996) argues by means of the raven paradox that considering not-Q elements to evaluate a universal claim is a very ineffective strategy in realistic situations. This does not mean that people do not search for exceptions, as he argues, but they do so by checking the elements that the statement is about: P cases. This is a relevance explanation of hypothesis testing, disregarding the logical requirement that not-Q items should be looked at. Second, the analogy between the 2–4–6 task and the selection task, made at the outset of this chapter, illustrates that the selection task standard is not universal. The hypothesis of the tester in the 2–4–6 task might be expressed as the statement: If I propose a triple with three even increasing numbers, the experimenter will give a "yes" answer. The original normative strategy here is not to check P and not-Q (this is impossible in the procedure of the task), but to check not-P. However, I have shown in Chapter 3 that checking P, even in the rule discovery task, is generally a better choice, because there we have the highest probability of finding exceptions (Klayman & Ha, 1987). Also, checking P is psychologically the most straightforward choice because it has highest psychological relevance, as Nickerson also shows. It is, in the perception of the tester, what the statement is about in the first place. Clearly, the selection task provides interesting new predictions and effects for hypothesis-testing behaviour, albeit precisely at the point where logical predictions fail.

CONCLUSIONS

The dilemmas of the application of statistical models to the selection task have been thoroughly discussed. The statistical assumptions made in these models are disputable from the perspective of logic. The assumed costs of test (card) choice (Klauer, 1999), probabilities of the terms in the statement, the interpretation of the hypotheses (the independence and dependence models) by the testers (Oaksford & Chater, 1994b), the utilities of decision errors (Kirby, 1994), are all fundamental in human hypothesis testing but basically irrelevant in conditional reasoning, because a conditional statement is about a relation between two elements that is strictly defined. Under the influence of the statistical approach, however, the selection task has evolved as a general hypothesis-testing task in which these statistical parameters play a major role. Oaksford, Chater, and Grainger (1999) argue that this task is to be considered, like every human reasoning task, as one of probabilistic reasoning. They even see logical solutions incidentally found in experiments as anomalies. I believe that this approach has indeed led to a significant contribution in understanding human hypothesis testing, especially the relevance explanation of numerous probabilistic and frequentistic effects, despite its persistence in following the rather complicated detour of conditional reasoning.

I shall now round off with two questions: How can one explain the selection task, and how can one explain human hypothesis testing? We have seen that many kinds of answers to the former have been proposed: defective truth tables, deontic reasoning, cognitive availability, relevance, mental models, and also set sizes. The most recent approach is the statistical-inference approach. Attempts to answer the first question may not have benefited from statistical theory as much as was hoped. The incompatibility between proposition logic, which was the breeding ground for the selection task, on the one hand, and the fundamentals of statistical-inference theory on the other, might have been underestimated. It makes modelling the participants' cognitions and behaviour by means of probabilities and utilities in this task difficult.

It is surprising that few alternative *logical* explanations of the selection task have been proposed. The discipline of formal reasoning has evolved dramatically in the last few decades, under the influence of similar developments in linguistics and computer science. Many new sophisticated logical systems have been developed, intended to enrich old traditional logical systems such as proposition logic. For example, logical semantics, intentional logic, dynamic logic, and the concept of "possible worlds", account for contextual influences in reasoning and subjective beliefs of the reasoner (Gamut, 1991; Montague, 1974; van Benthem, 1989). The fact that the most recent approach to selection task research originated in the world of statistical inference rather than that of logic might be due to historical practice. Cognitive psychologists are generally well trained in statistics but less so in logic. Also, the fundamental incompatibility between logic and empirical psychology, as Popper (1959/1974) stressed, may play a role. In this view, logic belongs to the realm of abstract truths, which has nothing to do with what people actually consider to be true. Wason attempted to bridge this gap, but not without Popper's disapproval (Wason, personal communication). However, looking around in modern logic might be a rewarding enterprise for future scientists in selection task research.

The second question was: What did these trends in selection task research contribute to the study of hypothesis-testing behaviour? Paradoxically, although the application of statistical theory in the selection task has been problematic and even confusing for the selection task itself, this new paradigm has contributed much to hypothesis-testing behaviour in general. Indeed, set-size effects, prior belief effects, more or less pragmatic utilities, and particularly relevance, have come to play their role in describing human-testing behaviour thanks to recent developments in selection task research. In the next chapter, I shall turn to hypothesis-testing experiments explicitly designed to be modelled by statistical-inference theories.

CHAPTER FIVE

Hypothesis testing under uncertainty

INTRODUCTION

In the selection task, the evaluation of the hypothesis tested by the tester can result in a demonstration of either its complete truth or its complete falsity, according to the standard of proposition logic. There is no intermediate conclusion possible, even if such a conclusion might be highly intuitive to an everyday reasoner. In most common hypothesis-testing situations, observations provide uncertain information about the hypotheses; they provide some support for a hypothesis, but no decisive support. When we test someone's personality by asking him the question, "do you like parties?", we cannot definitely conclude that he is socially awkward if he answers "no", but we do feel that we have gained some information about his character. The uncertainty assumption is an assumption made by statistical theories on testing (see Chapter 3). In these theories, tests themselves can be described in terms of the *amount* of support or counter-evidence they can give a hypothesis. The truth values of hypotheses can vary continuously from 100% certainty about their truth to 100% certainty about their falsity.

The present chapter presents hypothesis-testing research in which uncertainty is explicitly assumed to play a role, along with, occasionally, utilities and prior beliefs. First, some studies will be discussed in which tests are formulated in probabilistic terms. Then I shall deal with studies in which verbal testing is observed. The last ones mainly concern social hypothesis testing, i.e., situations in which social judgements are made by asking for information about people.

HYPOTHESIS TESTING WITH PROBABILISTIC TASKS

In tasks with probabilistic stimulus materials, the tester selects from various presented tests that are defined in terms of chances. In the majority of the experiments, the two terms of the likelihood ratio of a certain test result are given. The general pattern of these experiments can be described using a fictional example. The tester is asked to test a hypothesis, such as, "this person is an introvert". Subsequently, a number of possible tests are proposed. These are in the form of the likelihoods of confirming results, assuming the hypothesis H to be tested, along with the not-H alternative. In order to test the hypothesis that a person under scrutiny is an introvert, (H), the following tests would be proposed to the tester: Find out whether or not the person lives in a commune (X_1), never goes to parties (X_2), keeps a diary (X_3), and visits the pub every day (X_4). Sometimes, the alternative hypothesis is specified, sometimes it is not. The prior probabilities of H and not-H are given. The hypotheses are generally said to be equally likely at the outset. The tests (X) have two possible results: true ("yes") or false ("no"). This produces the matrix shown in Table 5.1 for this example, which is presented to the participant. The researcher defines the testing strategies by manipulating the chances in the likelihood ratio. The researcher observes the strategies followed by the tester, by observing the preference for tests with a certain likelihood ratio. For instance, a test with a greater chance of a confirming result for the focal hypothesis than for its alternative is called the confirming test in many studies. The tester is said to apply a confirming strategy when he or she prefers likelihood ratios with this feature.

Many variations of this pattern have been implemented. The great advantage of the probabilistic presentation of tests is that diverse testing strategies can be accurately defined and observed simply by manipulating the chances. The disadvantage is that the probabilistic representation of the possible tests is rather abstract, and strategies require statistical reasoning on the part of the testers. Due to this difficulty, the researcher is also uncertain as to whether or not the strategies defined have the same psychological significance for the tester as was intended to be measured. For example, the tester may prefer to select tests with high likelihood ratios. It remains dubious whether or not the participant's preference is actually caused by an intention to test "diagnostically".

In the course of the development of this experimental programme, the manipulations of the likelihoods in order to define testing strategies have become increasingly refined. Ultimately, the strategies defined converge with

TABLE 5.1
Example of a set of tests described as likelihoods, for testing the hypothesis H that "a person is introvert", presented to the participant in a typical probabilistic hypothesis-testing task

		p(yes | H)	*p(yes | not-H)*
X_1	lives in a commune	.30	.50
X_2	never attends parties	.80	.20
X_3	keeps a diary	.80	.50
X_4	visits the pub daily	.10	.40

the strategies defined in the previous paradigm: The following sections will deal with the testing strategies studied in the probabilistic paradigm.

Positive testing and the "confirmation" bias

Doherty, Mynatt, Tweney, and Schiavo (1979) had their testers discover on which of two islands a vase originated. The vase was found at the bottom of the sea. In a matrix similar to that mentioned previously, six features of the vase were described. Along with each feature, an indication was given of how often the feature occurred in vases from the first island (H) and in those from the second island (not-H). The likelihoods, 12 altogether, were covered with stickers. Based on observations of the vase, the testers first formulated a hypothesis concerning its origin. Doherty et al. found two testing tendencies in the testers. First, the testers were inclined to confirm: a confirmation bias. Second, the likelihood regarding the alternative hypothesis was neglected. The second finding is discussed in the next section. Confirmation bias was expressed as a preference for information about likelihoods relating to one's own hypothesis only. A tester who assumed that the vase originated on the first island (a proponent of H) preferred to examine the likelihood of certain features of the vase occurring on the postulated island of origin (one's own hypothesis).

A related tendency, which also came to be known as confirmation bias, has been found in a number of studies in which all likelihoods are presented to participants (Bassok & Trope, 1984; Beyth-Marom & Fischhoff, 1983; Trope & Bassok, 1982). In these studies, confirmation bias was regarded as being a preference for tests in which the likelihood of a positive result under the focal hypothesis is larger than under the alternative hypothesis. In our example concerning introversion, this means a preference for testing the X_2 predictions (the person under scrutiny never attends parties) and X_3 (the person keeps a diary), as displayed by the likelihoods in the stimulus materials.

However, this latter variant of confirmation bias can actually be shown to be a positive testing strategy (Skov & Sherman, 1986). A positive test is a test in which a possible positive result (a "yes" answer) confirms the focal hypothesis, but which does not necessarily lead to such a result (see Chapter 3). When the likelihood of a positive result under the hypothesis being tested is greater than under the alternative, means that, if this result occurs, the tester will interpret it as a confirmation of the hypothesis. Imagine that the (confirming) predictions in the example, X_2 and X_3, turn out to be true (the person never attends parties and he does keep a diary); the tester will then assume that the introversion hypothesis is correct. But the answer can be "no" and hence the hypothesis is falsified. This definition is conceptually identical to the confirmation bias that Wason (1960) also found in the rule discovery task. Similarly, a tester cannot force confirmation using this strategy: Choosing X_2 and X_3 does not ensure that the person under scrutiny will actually respond with a "yes".

Diagnosticity and pseudo-diagnosticity

The second testing strategy reported by Doherty et al. was that testers neglected to choose the likelihoods of one feature occurring both under one's own hypothesis and under the alternative. The testers only asked about the probability of a certain feature of the vase occurring on the postulated island of origin. They did not choose pairs in the matrix, although both chances (from the right- and the left-hand column) are necessary to formulate the likelihood ratio of a test. Participants were of the opinion that sufficient information could be obtained from the first sort of chance alone. Doherty et al. called this second sort of testing strategy "pseudo-diagnosticity". Pseudo-diagnosticity has been found in a number of experiments in which the tester, just as in the case of Doherty et al., must indicate which information is regarded as important for testing the hypothesis (Doherty & Mynatt, 1990; Mynatt, Doherty, & Sullivan, 1991). This strategy has also been demonstrated in the rule discovery tasks (Gadenne & Oswald, 1986). The combination of confirmation bias *and* pseudo-diagnosticity as defined by Doherty et al. in their experiment can be summarised as the tendency of testers testing H to prefer high likelihoods under H and to underestimate the importance of the denominator in the likelihood ratio (the likelihood under not-H). The formulations of "pseudo-diagnosticity" in the Doherty et al. studies are distinguished from the standard pattern in that the testers were not directly presented with the chances. Indeed, the likelihoods were covered with stickers.

In a similar study, Mynatt, Doherty, and Dragan (1993) replicated the pseudo-diagnosticity effect. In this study, the likelihood $p(x_1 | H_1)$ was given. Participants could ask for one more piece of information; either $p(x_1 | H_2)$,

$p(x_2 | H_1)$, or $p(x_2 | H_2)$. The test situation was about two mutually exclusive hypotheses, and two possible outcomes of the test. The authors found that when the test situation was about two actions (H_1 and H_2) rather than between accepting (H_1) or rejecting (H_2) the hypothesis, the pseudo-diagnosticity effect disappeared. The participants chose information about the alternative option ($(x_1 | H_2)$). Interestingly, the "bias" also disappeared when the likelihood displayed was lower than .50, i.e., when the evidence favoured the alternative hypothesis. The authors explain this effect mainly in terms of working memory capacity. People can have just one hypothesis at a time in this working memory. Which hypothesis they consider depends on relevance effects. The phenomenon is basically the same as that observed in the selection task. When a hypothesis is unlikely *a priori*, or at least has exceptions, the tester tends to look for the falsifying cases. Apparently the untrue character of the hypothesis directs our attention to the cases that actually make it untrue (see also Evans & Over, 1996).

With regard to the effect of the action context, the authors suggest that people might have an intuitive understanding of utility maximisation but not of diagnosticity because of the impossibility of considering two hypotheses at a time. A problem for this explanation is that in tasks in which both likelihoods are given, people actually are sensitive to diagnosticity, as will be shown below. An important psychological difference between the two tasks might be that in the action problem there are two equivalent options among which no choice is made yet: Indeed, the participant must choose which of two cars to buy. This might have distributed attention over both possible cars. The facilitation of the action context can thus be interpreted as a form of practical reasoning that, in the selection task, occasionally causes a higher activation of the alternative hypothesis (Manktelow, 1999). In contrast to Mynatt et al.'s hypothesis, people might, in this context keep two hypotheses in mind. In the "inference" version, one of the two hypotheses (cars) was the true one. In this problem, people might rather represent one car as a "working hypothesis". In other words, the context might have facilitated the representation of one rather than two hypotheses depending on what is perceived as relevant in a context. The hypotheses getting relevance are the one(s) for which the likelihood seems relevant. This would also be consistent with the finding that people are actually sensitive to diagnosticity when both likelihoods are explicitly given. Differences between pure testing situations and action choices need more investigation in the future, however.

When the likelihoods in the matrix are presented, testers do appreciate the diagnosticity of a test, as can be seen in the following series of experiments. In these, the diagnostic strategy is defined as a preference for tests that are capable of distinguishing, as much as possible, between H and not-H. These are defined as tests with a high expected diagnosticity (Bassok &

Trope, 1984; Beyth-Marom & Fischhoff, 1983; Skov & Sherman, 1986; Trope & Bassok, 1982). There are several ways to calculate this capacity of a test. An adequate measure is the expected log-likelihood ratio, which was previously discussed in Chapter 2 and Chapter 4. In many studies of the present type, a simplified version of it is used. Typically, Skov and Sherman's (1986) simple formula is used in experimental studies that have the task structure of the example. They calculate the expected diagnosticity of a test as follows:

$$p(\text{“yes”}) \cdot \frac{p(\text{“yes”} \mid H)}{p(\text{“yes”} \mid \text{not-}H)} + p(\text{“no”}) \cdot \frac{p(\text{“no”} \mid H)}{p(\text{“no”} \mid \text{not-}H)} \qquad (5.1)$$

Mostly, a correction to this measure is applied: In calculating the diagnosticity, the numerator and the denominator of the likelihood ratios of the possible test results must be taken in such a way that the numerator is greater than the denominator. The correction ensures that a test with a likelihood ratio of .90/.50 has the same diagnosticity as a test with a likelihood ratio of .50/.90. In other words, the diagnosticity thus becomes, regardless of what one regards as the focal hypothesis and what one regards as the alternative(s), a measure of the capacity to discriminate between the hypotheses (Bassok & Trope, 1984; Skov & Sherman, 1986). This emphasises that it is a symmetrical measure for the quality of a test (see Chapter 4). Thus, the expected diagnosticity is the product of the chance of a test result, multiplied by its likelihood ratio, and summed over all possible test results. In our example, X_2 is a relatively diagnostic test.

However, diagnosticity can be approximated with an even simpler measure: the difference between the likelihoods of the numerator and the denominator (Baron, 1985; Klayman, 1995). From a psychological point of view, it is quite plausible that people globally and quickly estimate the diagnosticity of a test in this way. Testers do indeed prefer tests with highly differing likelihoods (Bassok & Trope, 1984; Beyth-Marom & Fischhoff, 1983; Skov & Sherman, 1986; Slowiaczek et al., 1992). Summarising people's sensitivity to diagnosticity testers do not *spontaneously* take into account the extent to which a test result supports a theory other than the focal one. But they do prefer diagnostic tests when all the information about the likelihoods is available.

At this point, two phenomena of testing behaviour in the probabilistic testing tasks have been discussed. First, people prefer tests with a higher likelihood of a particular observation being true under the focal hypothesis than under the alternative. This is basically equivalent to what was called "positive testing" in the rule discovery task. Second, people prefer diagnostic tests when all the information needed to assess this diagnosticity is available to them. At this point, we come back to the discrepancy between the insensitivity to diagnosticity found in the *test* selection paradigm and the sensitivity to diagnosticity found in the *likelihood* selection paradigm. Mynatt et al.'s (1993) explanation is that people do not select information analytically, but on the basis of perceived relevance. In contrast, once the information is available they use it analytically. This explanation is in line with Evans and Over's (1996a) distinction between heuristic and analytic stages of reasoning.

But the contrast between the findings in the two experiments might not be that large when we consider that testers having all the information (test selection paradigm) also prefer *positive* tests, in addition to diagnostic tests. This might be due to the tester's attention being directed at the focal hypothesis. This, in turn, might also be due to relevance judgements. Moreover, it provides an explanation for the suppression of the pseudo-diagnosticity effect, when the tester expects the evidence (likelihood) that she selects to favour the alternative hypothesis. This result was found by Mynatt et al. (1993) in their likelihood selection experiments: Attention was directed at the hypothesis for which the likelihood was "positive". In sum, the pseudo-diagnosticity findings can be seen as an expression of the same process causing the positivity effect: The evidence is preferred that has highest likelihood under the hypothesis under focus. And, as I have shown, the positivity effect is not an analytic and strategic tendency to have one's hypothesis confirmed. It might rather be a reflexive response to see positive tests as more relevant.

Thus, these findings show that people are not demonstrably liable to a real confirmation bias, in the sense of the focal hypothesis being allocated a higher chance of surviving the tests than the other hypothesis, due to some strategic behaviour. In the following section, I shall show that actually demonstrating such a bias requires taking into account the way participants interpret test results at two stages: after the test has actually been performed, and in their anticipated interpretation of the results.

Combinations of test selection and outcome interpretation

Recently, hypothesis-testing studies with probabilistic materials have been performed in which confirming strategies other than positive testing have been investigated. The interesting question here is whether or not it is possible to define a testing strategy in which confirmation of the hypothesis *can* be forced or at least facilitated. A preference for this strategy would indicate something that could justifiably be called a confirmation bias in testing behaviour. This same question emerged in analyses of the rule discovery task. In the beginning of this programme, positive testing was explained by confirmation tendencies. However, theoretical and empirical arguments refuted this (see Chapter 3). In the present programme, this question was first addressed by Skov and Sherman (1986) and Slowiaczek et al. (1992). I shall elucidate their strategy in the light of our example. It will be referred to as the "real confirmation bias".

The tester wishing to formulate the test in such a way that it promotes confirmation of the hypothesis must maximise the chance of a result that supports the hypothesis. In the example given at the beginning of this chapter, which test results will lead to confirmation of H (the introversion hypothesis)? In the case of the X_1 test, it is the answer "no" (the person under scrutiny does not live in a commune). In the case of X_2 and X_3 it is "yes", and for X_4 it is "no". What is the prior probability that these results will turn up? Assume that the tester does not know in advance whether the introversion hypothesis (H) or its alternative is correct. Both are *a priori* equally likely. This is also assumed in most experiments: (p(H) = p(not-H) = .50). The chances of a confirming result from each test in the example are:

$$X_1 \ p(\text{no}) = p(\text{no} \mid \text{H}).p(\text{H}) + p(\text{no} \mid \text{not-H}).p(\text{not-H}) = .60$$
$$X_2 \ p(\text{yes}) = p(\text{yes} \mid \text{H}).p(\text{H}) + p(\text{yes} \mid \text{not-H}).p(\text{not-H}) = .50$$
$$X_3 \ p(\text{yes}) = p(\text{yes} \mid \text{H}).p(\text{H}) + p(\text{yes} \mid \text{not-H}).p(\text{not-H}) = .65$$
$$X_4 \ p(\text{no}) = p(\text{no} \mid \text{H}).p(\text{H}) + p(\text{no} \mid \text{not-H}).p(\text{not-H}) = .75$$

The tester wishing to confirm selects the test in which the chance of a confirming result is high, such as, for example, the X_1, X_3, or X_4 tests. Now, the test choice alone ensures that the hypothesis of the tester has a greater chance of being confirmed, whatever the true state of nature. The demands that the tests must fulfil, according to Skov and Sherman and Slowiaczek et al., in order to cause this "real confirmation bias" can be summarised in the following criterion: The chance of a confirming result c (notice that this can be either "yes" or "no"), of the test of the hypothesis H, must be greater than the *a priori* chance of this hypothesis actually being true. Thus, the test leads to a confirmation bias if $p(c) > p(H)$ is valid. In the example, this applies to X_1, X_3, and X_4.

Skov and Sherman and Slowiaczek et al. presented testers with tests of this sort. Testers had to imagine that they had landed on a strange planet where there are two sorts of creatures (they are called "Gloms" and "Fizos" here, as in the experiment performed by Slowiaczek et al.). There are just as many Gloms as there are Fizos (this is information concerning the priors). These beings are invisible. The testers encounter a being and have to discover whether it is a Glom (or a Fizo, in another condition). The testers may ask two questions (perform tests). They receive a matrix indicating the likelihoods of occurrence of 12 features (of the Gloms and Fizos) as in the standard pattern. They then select two features. The matrix with the chances is formulated in such a way that three dimensions of testing behaviour can be examined: the positive testing strategy, the diagnostic strategy, and the real confirmation bias. Skov and Sherman and Slowiaczek et al. found that testers displayed all three strategies: They preferred the positive tests and diagnostic tests and displayed, in addition, a real confirmation bias. That is, they preferred tests with a higher probability of a supporting result than they believed the hypothesis to be true, *a priori*.

Interestingly, such a preference has consequences for the *values* of the possible test results. The more that test selection is subject to this bias, the smaller its evidential value for supporting the hypothesis, as was shown in the previous analysis of the philosophical and formal theories of testing. Hence, the interpretation of the result by the tester is important in interpreting his or her testing strategy. The relation can be illustrated with reference to our example. According to Bayesian reasoning, one should attach less certainty to a confirmation of a hypothesis than to a refutation, if one performs a test in which the probability of confirmation is maximised. The confidence that a tester should have in the hypothesis as a result of the test outcome can be calculated as follows: Suppose the tester chooses X_3 with a likelihood ratio of .80/.50, thus with a .65 chance of a hypothesis-supporting result ("yes, the person keeps a diary"). The test indeed turns out to be positive. According to Bayes' theorem (see Chapter 2), the posterior probabilities are:

$$\frac{.80}{.50} \cdot \frac{.50}{.50} = \frac{8}{5}$$

After this confirming result, the tester should place approximatley one-and-a-half times (8/5) as much faith in the probability of the focal hypothesis H as in the probability of the alternative. But suppose that the outcome is negative: the person under scrutiny does not keep a diary:

$$\frac{.20}{.50} \cdot \frac{.50}{.50} = \frac{2}{5}$$

The calculation indicates that the tester can generally place two-and-a-half times (2/5) as much faith in the *alternative hypothesis* as in his or her own hypothesis. Thus, falsification leads to a stronger revision of belief than confirmation (see Chapter 3).

Now, the tester can take this asymmetry into account when considering the test. Skov and Sherman and Slowiaczek et al. studied the extent to which testers bring test selection into line with test result interpretation. Participants with a biased test selection were asked to assign subjective confidence to the hypothesis after a positive and after a negative result. These testers had, however, the same amount of confidence in the hypothesis after a positive result as after a negative result. They were thus overconfident after a confirming result and underconfident after a refuting result. Slowiaczek et al. refer to this as the "equal and opposite assumption". They mean that the tester assigns just as much value to one result as to the opposite one. Summarising, the "real confirmation bias" is a strategy that really allows the manipulation of the probability of a supporting test outcome. Moreover, it influences test outcome value, *a posteriori*. The strategy was observed in two experiments. In these, people did indeed prefer to maximise the probability of a confirmation. Also, they neglected the normative consequences this should have on their evaluation of the result.

Probabilities versus evidential value of outcomes: The probability value model

The "real confirmation bias" has many properties in common with the theoretical analyses of testing strategies already discussed. Those similarities will be dealt with briefly here. Klayman (1995) stresses that the bias discovered by Skov and Sherman and Slowiaczek et al. arises from the *combination* of a test preference and the interpretation of the results of these preferred tests. This test preference can be described as a preference for checking those data that are either very probable or very improbable under the assumption that the favoured hypothesis is true (extreme likelihoods for the favoured hypothesis), whereas, for the likelihoods of the non-favoured hypothesis, no extreme values are chosen. In Figure 2.1 (Chapter 2), the effect of extreme likelihoods on the information value of the result was displayed. Extremity differences between the likelihoods lead to proportionate differences in the information values of the results. Participants tend to extremity with respect to the likelihood term regarding their hypothesis. But extremity in test selection *per se* does not lead to a testing bias. It only does so if, after the test has been performed, the tester does not adapt his or her belief revision in accordance with the statistical consequence of this preference. Indeed, as shown above, an observed confirming outcome with a relatively extreme likelihood provides weak support for the hypothesis it favours. If the participant takes into account the low value of this support, nothing goes wrong.

Slowiaczek et al. showed that testers are not aware of the relation between probability and value in the task set-up. This unawareness can occasionally lead to biases. The question arises, why? Klayman (1995) denies that this unawareness is motivational. He argues that the calculations necessary to bias hypothesis testing in such a way as to favour supporting test results are very complicated. Therefore, this bias is unlikely to be the product of a conscious strategy. The bias is probably due to cognitive constraints. This, I think, is certainly true for tests formulated in terms of quantitative likelihoods, as in the experiments of Slowiaczek et al. and Skov and Sherman; but what about the "real confirmation bias" in more realistic situations with conceptually defined tests?

Poletiek and Berndsen (2000) propose the following analysis for more realistic situations of testing. The hypothesis tester selecting a test makes a pre-posterior analysis of several possible tests in which the probabilities of different outcomes are considered along with the evidential values of these outcomes for demonstrating the hypothesis to be true (see also Poletiek, 1995). We assumed that the inverse relation between the probability of a confirmation and its supporting value is intuitive in most realistic situations, especially in situations in which the tests are described conceptually rather than numerically. Consequently, testers can manipulate the probability of a confirming outcome and its value by selecting tests to satisfy their goals. Hypothesis testing and its dilemmas are, according to Poletiek and Berndsen, analogous to choosing a game in gambling theory: Either the probability of winning is high but the value of the prize is low, or the probability of winning is low and the prize is high. Now, if we assume that a tester favours one hypothesis, he or she may be motivated to adopt a confirmation bias in two ways: Either the *probability* of obtaining a confirmation can be maximised (as occurred with Slowiaczek et al.'s participants) at the cost of its *value*, or the *value* of a confirmation can be maximised, at the cost of its *probability*. Poletiek and Berndsen illustrate this model intuitively as follows.

Suppose that there is the hypothesis that Mary has stolen a ring at the jeweller's. This hypothesis is linked to two predictions. Test A consists of looking for the ring in Mary's house. Test B involves investigating whether or not Mary was seen on the day of the theft in the section of town where the jeweller's shop is situated. In this case, test A has a possible confirming outcome that is less probable *a priori* than test B. Indeed, if we know nothing about Mary being suspected of theft, we would intuitively guess that the probability that Mary possesses the specific ring is a great deal lower than the probability that she was in the jeweller's area that day. Which test should be carried out? Checking the first prediction is risky with regard to the "Mary-is-guilty" hypothesis. The probability of a confirmation (finding the ring) is low; the risk of refutation is high. But if the ring is found, the proof of Mary's guilt is strong. Checking the second prediction is a less "risky" test; the risk of rejection is lower. But if she was indeed seen in this area of town, the proof of Mary's guilt is still weak.

Both tests are advantageous in one respect and unfavourable in another. The tester makes a trade-off. The important thing in the probability value model proposed, is that while the relation between the prior characteristics of a test and its implications for the values of the outcomes is indeed rather complex to describe formally, it is also very intuitive in many real-life test situations. In general, authors assume that participants merely tend to maximise the probability of a confirmation of a desirable hypothesis (Trope & Liberman, 1996). However, this disregards the fact that this confirmation bias implies a willingness to accept the favourite hypothesis on the basis of poor value evidence. And it is questionable whether or not people who lean towards one hypothesis and are aware of this implication will indeed always prefer to sacrifice information value for the benefit of a greater chance of acceptance. There are several interesting examples in the research dealt with above, in which the tester was found to minimise the probability of false confirmations, when he or she strongly believes the hypothesis to be true. For example, in the context of a decision making model of selection task card preferences, Klauer (1999) calculated on the basis of his non-sequential model (see Chapter 4) the utilities participants assign to different kinds of errors, and found that the most avoided error when the hypothesis was favoured was false acceptance.

Thus, it is hypothesised that, in some realistic test situations, testers might maximise the *value* of a confirmation of the hypothesis they favour, instead of its probability, precisely minimising the probability of confirmation. This was found in one analysis of the selection task. This is also what Poletiek and Berndsen (2000) found in another paradigm: Participants do indeed tend to maximise confirming outcome value at the cost of its probability in some realistic contexts. The tasks arranged by Poletiek and Berndsen involved a scientific and a juridical context. In neither of the contexts did participants tend to maximise the probability of support for the focal hypothesis. In the juridical context, they even tended to maximise the value of the evidence, at high risk of rejection of the desirable hypothesis. Thus, when participants had to choose a test for finding out whether or not a defendant was guilty, they preferred a test that could possibly provide strong incriminating evidence for guilt, especially when the defendant's acts were highly disapproved of. In these cases, participants were motivated to have the "guilt" hypothesis confirmed. Poletiek and Berndsen (2000) show that balancing the two test characteristics is equivalent to setting the tolerated risks of decision error (see also Friedrich, 1993).

In sum, a real confirming strategy is complex to describe formally. It has two stages: test preference and interpretation of the outcome. These two characteristics are related in a way that is fairly complex to calculate when the tests are described in terms of likelihoods. This might be why participants fail to interpret the test results according to their formal relation to the chosen test, when faced with formally described tests. However, this relation is intuitive in many everyday test situations. When it is, the "real confirmation bias" (maximising the prior probability of a confirming result) can disappear, even for desirable hypotheses, and may be replaced occasionally by a strategy that aims at obtaining the highest value result (maximising the posterior value of a confirming result).

Information bias

Another "probabilistic" strategy of testing was first described in probabilistic experiments by Baron, Beattie, and Hershey (1988). They studied a new testing bias, "information bias", within a Bayesian framework. The study is interesting because it contrasts with other probabilistic experiments. For example, it does not question confirmation bias effects but it questions whether or not testers use useless tests, i.e., tests that are expected to add nothing to what is already known. In contrast to the previous experiments, the Baron et al. tasks present participants with more than two hypotheses to be tested. Moreover, these hypotheses have different prior probabilities. Broadly speaking, Baron et al. proceed from the point of view that a test is more useful when the probability of the correct hypothesis being adopted increases. The information bias then amounts to the tester performing a "useless" test in this sense. The information bias can be illustrated using one of their examples.

In the construction created by Baron et al., different hypotheses about medical diagnoses are presented with prior probabilities, along with a series of tests, according to the standard pattern. The tester is informed that a test has to be carried out on a patient. This patient suffers from one of three possible illnesses (H_1, H_2, or H_3). The tests are medical tests. The likelihood of a positive result under the three hypotheses is given for each test. Table 5.2 shows some of the data as presented to the participants.

How can the usefulness of a test be established? The H_1 illness is initially regarded as being the most likely. Without a test, it is decided that H_1 actually is the illness. Thus, the issue is about performing a test which could possibly lead to a conclusion other than accepting H_1. This is, therefore, a test that could provide a tester with more information than is available in the initial situation. Baron et al. (1988) constructed the following model for the usefulness of a test. In the first place the "joint probability" of the likelihood of a certain test result and the prior probability of the hypothesis are calculated. This is done for all possible results of a test. In the materials of Table 5.2, these are a positive and a negative result. The highest joint probability for the positive result is added to the highest joint probability of the negative result. This sum represents the expected impact of a test in deciding about the hypotheses on top of what is already known about these hypotheses. If this sum is larger than the chance of the most probable (prior) hypothesis, the test is regarded as being useful.

This only applies to Test 3 in Table 5.2. After all, whatever the result of the first test (Test 1), one would also opt for the most probable hypothesis (H_1). The "joint probabilities" of each possible outcome are largest for H_1. The sum is $p(+ \mid H_1) \times p(H_1) + p(- \mid H_1) \times p(H_1) = (.50 \times .64) + (.50 \times .64)$. This sum is equal to .64 which is the chance of the most likely hypothesis.

TABLE 5.2
Tests with their likelihoods of a positive result, for testing three hypotheses, as presented to the participant by Baron et al. in a typical probabilistic hypothesis-testing task

	$p(+\mid H_1)$	$p(+\mid H_2)$	$p(+\mid H_3)$
Test 1	.50	1.00	.00
Test 2	.25	.00	.00
Test 3	.00	1.00	.00

$p(H_1) = .64$, $p(H_2) = .24$, $p(H_3) = .12$.

Thus, we always accept H_1, either with or without knowing the result of Test 1. This also applies to Test 2. Test 3 is different. As Table 5.2 indicates, H_2 will be accepted in the case of a positive result, since this result excludes both of the other hypotheses. The usefulness of this test can be calculated from the Baron et al. model: The sum of the largest "joint probabilities" for the positive and negative result is: $p(+\mid H_2) \times p(H_2) + p(-\mid H_1) \times p(H_1) = (1 \times .24) + (1 \times .64)$. This is equal to .88, exceeding the chance of acceptance of the prior most probable hypothesis. Baron et al. presented their testers with 10 tests according to the example. Eight of these were useless and two were informative. The testers had to indicate how relevant they found the tests. Baron et al. found that the testers displayed the "information bias": They indicated that they found both informative and uninformative (useless) tests relevant.

In contrast to the probabilistic experiments executed by Skov and Sherman and Slowiaczek et al., the exact values of the posterior probabilities of the initial hypotheses are not relevant for Baron et al. Although their model is based on Bayesian calculations, they identify hypothesis testing with a discrete decision problem: Accept one hypothesis and reject the others. However, such revisions of belief can indeed be calculated in the materials of Baron et al. For example, a positive test result of Test 2 would cause some decrease in belief in hypothesis H_1, and totally exclude hypotheses H_2 and H_3. This information would not lead to another decision about what is true, but the revisions of belief, especially the total exclusion of some options, compared to the small increase in probability of other options, might be relevant in real-life situations. Thus, in contrast to Baron et al.'s assumptions, Test 2 may be a good choice in some practical situations.

In the following section, I turn to testing behaviour experiments without chances or numbers. The tasks have verbal materials, mostly relating to social hypothesis testing. In the context of social judgement, social interaction processes add up to the phenomena discussed previously, and can result in testing biases. For example, the very formulation of a question can cause a respondent to prefer a certain answer, irrespective of the true answer. However, even in this "ecological" paradigm, the concepts of confirmation bias, positive testing, and diagnosticity are the basic features, hence relating it surprisingly well to the previous hypothesis-testing research.

TESTING WITH VERBAL TESTS

Snyder and Swann (1978) presented their testers with descriptions of an introvert or extravert person. The tester then had to judge another person in terms of the similarity to the described character. Thus, they tested the hypothesis, "this person is introvert/extravert". To do this, they were presented with a number of questions, from which they could select several, to ask the person under scrutiny. The strategies of hypothesis testing could be categorised by the sort of question that the tester selected: Some questions were defined as being "extravert", others as "introvert", and others as "neutral". According to the authors, extravert questions were those that one would ask someone who is already regarded as being extravert (for instance, "what would you do to liven up a party?"). Introvert questions are those that one would ask someone who is already regarded as introvert (for instance, "what do you find unpleasant at noisy parties?"). The neutral questions were those where there was no agreement about their classification. Snyder and Swann found a very strong effect of the nature of the hypothesis being examined upon the nature of the questions asked. If the person to be judged was introduced as being extravert, he was almost exclusively asked "extravert" questions. The authors conclude that the testers displayed a strong tendency to use a confirming test strategy when testing hypotheses about the social characteristics of people.

Positive testing and confirmation bias

The "confirming" testing strategy found by Snyder and Swann has, as an inherent feature, the presupposition that the trait that is to be judged is already present in the person under scrutiny. Swann and Giuliano (1987) however, refer to such questions as "constraining questions". These questions force the person under scrutiny to give a certain answer. The extravert question, "what would you do to liven up a party?" facilitates an answer that will make the respondent appear extravert. Swann and Giuliano separated the "constraining" from the "confirming" strategy. They defined confirming testing in yet another different way.

In their first experiment, testers had to think up questions in order to find out whether the person under scrutiny conformed to an introvert or extravert character profile. In contrast to the experiments of Snyder and Swann

(1978), a question was regarded as being confirming when a "yes" answer indicated the presence of the trait at issue (such as, in the evaluation of extraversion, "do you talk a lot at parties?"), and as being disconfirming when a "yes" answer indicated that the person under scrutiny possessed the opposite character trait (such as: "Are you generally quiet in a group?"). Swann and Giuliano found that testers had a spontaneous preference for confirming questions. In the second experiment, the testers were divided between two conditions, according to the degree of belief in the hypothesis that the person under scrutiny actually did possess the character trait: a certain and an uncertain condition. The questions were divided over two dimensions: confirming versus disconfirming questions and constraining versus non-constraining questions. An example of a confirming but non-constraining question for extraversion was, "are you inclined to liven up parties?". An example of a constraining non-confirming question was, "How would you liven up a party?". Swann and Giuliano found that testers posed more confirming *and* constraining questions when they had stronger prior beliefs about the hypothesis. I shall return to constraining tests at the end of this section.

In the third experiment, testers were again presented with a number of confirming and disconfirming questions (in line with the definition in the second experiment) and were asked to indicate how "informative" they would find the responses to these questions. The intention was to examine which sort of questions testers would see as being "informative". Testers appeared to find confirming questions the most informative. According to the researchers, it may be concluded that testers do not display any insight into the informativeness of questions but, in contrast, are liable to a confirmation bias when testing social hypotheses.

Interestingly, Swann and Giuliano's (1987) confirming strategy is basically a variant of the positive testing strategy. The strategy is confirming, according to their definition, if a positive result (a "yes" answer) confirms the hypothesis being tested. The objection to this strategy, being a tendency to confirm, has already been elucidated by, among others, Wason's rule discovery task: Confirmation cannot be enforced using the positive strategy. The designation of this strategy, nevertheless, arouses the impression that the tester is aiming at confirmation of the hypothesis. In contrast, constraining questions do cause a bias in the long run, that is, a tendency to attribute too much confidence to the focal hypothesis due to elicitation of hypothesis-confirming answers.

Another experiment, in which the tests are verbally presented, was performed by Tschirgi (1980). Here, the tests are descriptions of small experiments instead of questions. The testers are presented with a number of small stories. One story, for instance, tells of someone who bakes a cake. He uses three ingredients: honey, wholemeal flour, and margarine. The cake turns out either successfully (in the "positive result" condition) or disastrously (in the "negative result" condition). Subsequently, the testers are to test the hypothesis about the cause of the end-product. In the case of the cake, this is: "The person thought that the cake turned out successfully/ disastrously due to the honey". Next, the tester may choose one of the three experiments presented. The three experiments represent three different testing strategies and consist of baking the cake anew, but with variations this time. In the first experiment, two elements are kept constant (margarine and wholemeal flour) and the crucial variable is altered (sugar instead of honey). This is the disconfirming strategy according to Tschirgi. In the second experiment presented, the crucial variable in the hypothesis is kept constant (honey is used) and the others are varied. This is the confirming theory. In the third experiment, all elements are varied. Tschirgi expected that the tester would follow the disconfirming test strategy when the hypothetical crucial variable had led to a negative result. In the case of a desired result, Tschirgi expected a preference for the experiment in which the crucial variable was kept constant (the second experiment). Tschirgi found the expected link: One chose "disconfirming" experiments significantly more often when one had to test a hypothesis concerning the cause of an undesired event, and a confirming experiment when the expected result was more pleasing to the tester (Tschirgi, 1980, p. 8).

> When an outcome is negative, children and adults test their hypotheses about the cause of the outcome in a sensible and logical fashion to produce disconfirming evidence. . . . In contrast to this logical behaviour in bad-outcome stories, adults and children systematically seek confirming evidence in good-outcome stories.

Did Tschirgi's participants seek confirmation? Tschirgi's confirmation bias might be regarded as merely a positive testing strategy. After all, in line with Swann and Giuliano's definition (1987), Tschirgi's confirmation test leads to a confirmation of the hypothesis only *if* it comes true. However, if the cake is again successful (or unsuccessful) in the experiment in which the honey has been omitted, this is a *refutation* of the hypothesis that the honey was the cause of the result. It is obvious that even when the confirmation experiment is used, the cake can be a disaster. If participants subsequently interpret the disaster as a falsification of the honey hypothesis, one cannot say that there is a confirmation bias here.

In sum, the uneasy relation between confirmation bias and positive testing emerges in hypothesis-testing studies with verbal materials as well. The pervasive positive testing in verbal materials experiments can possibly best be explained by means of relevance judgements. Testers tend to design tests for their guesses that are directed at the very concepts that figure in the hypothesis (see also Chapter 3). But this is not essentially wrong, as was argued above.

However, another characteristic of social hypothesis testing is asking "constraining" questions, which is a tendency that seems less "innocent" than positive testing. The troubles are caused by a combination of preference for these kinds of tests and the response behaviour of the respondent. This tendency has also been described as the "acquiescence bias" (Zuckerman, Knee, Hodgins, & Miyake, 1995). People testing hypotheses by asking others questions tend to formulate their questions in such a way that the hypothesis is "framed" in the question, as in Snyder and Swann's study. These questions elicit confirmations of the framed hypotheses. Thus, respondents tend to answer "yes" anyway. The combined effect of positive-test preference and acquiescence bias is a confirmation bias leading to confirming repeatedly prior social beliefs.

Another source of possible true confirmation bias tendencies is often associated with hypothesis testing, and arises from the evaluation of ambiguous evidence. People tend to attribute more credibility to evidence that supports their prior belief, when the available evidence is ambiguous. Lord, Ross, and Lepper (1979) presented their participants with both supporting and counter-evidence about the efficacy of capital punishment. They found that participants weighted more significantly and were more influenced by evidence consistent with their prior belief, and this influence took the form of strengthening existing beliefs. A similar phenomenon was found by Koehler (1993) among scientists. Scientists were more critical about the methodology of those studies disputing their own theoretical beliefs than about those whose results they believed to be true. This bias, however, is not caused by *testing* behaviour, because the evidence is not sought but is already available. It is mentioned here because it is frequently associated with hypothesis-testing bias. Confirmation bias as evidence-interpretation bias is very well documented (see, for an overview, Trope & Liberman, 1996).

Accuracy and utilities in social hypothesis testing

Finally, I want to discuss some recent theoretical approaches to hypothesis testing under uncertainty within the decision making framework. In this view, hypothesis testers are assumed to choose their tests according to their motivation to reduce a certain type of error (Friedrich, 1993; Trope & Liberman, 1996).

Trope and Liberman (1996) make a distinction between symmetric and asymmetric error reduction in hypothesis testing. Notice that this parallels my distinction between symmetric and asymmetric testing in the selection task (Chapter 4). Symmetrically reducing decision errors means equally reducing them both regarding the focal hypothesis and the alternative. This happens when testers are driven by a motivation for accuracy (Kruglanski, 1996). When the tester wants to reach a quick conclusion, he or she will lack this motivation. Asymmetrical motivation leads to what is classically called "biased testing": tending to prevent the focal hypothesis being rejected (Friedrich, 1993). People are generally presumed to do this by avoiding false negatives (Trope & Liberman, 1996, p. 265).

> We have argued that in testing a desirable hypothesis, it is often more important to individuals to avoid rejection than to avoid false acceptance.

Trope and Liberman, however, emphasise that testers can be concerned with both symmetrical and asymmetrical motives. They might want to make as few errors as possible but, from the errors they cannot further minimise, they might tolerate more of one type than the other.

An important feature of Trope and Liberman's analysis is the assumption that the costs of error presumably determine the levels of confidence that individuals require in order to accept or reject a hypothesis. As the citation makes clear, Trope and Liberman presume that testers generally demand low confidence to accept a desirable hypothesis to be true. However, Poletiek and Berndsen deny that this level of confidence is necessarily minimised for a favoured hypothesis, as I argued previously. They propose that testers balance the required level of confidence for accepting the favoured hypothesis against its probability of being accepted. Favouring one hypothesis might be psychologically equal to perceiving false rejection as a high cost, but also to regarding false acceptance as a high cost. Being biased *a priori* towards one hypothesis might induce a motivation to provide the best possible proof of it, as Poletiek and Berndsen found, rather than "as many proofs" as possible. Thus, both Trope and Liberman and Poletiek and Berndsen use the same theoretical pragmatic and subjective framework in describing hypothesis testing, but come to different predictions as to how people might "bias" the test procedure when one hypothesis is more desirable than another. There is some evidence for both points of view, but they need to be investigated further in future research, pitting them against each other in order to elucidate the conditions under which the confidence level or probability are maximised. Social hypothesis-testing contexts or other contexts with a "rich" content are especially interesting to investigate with regard to this question.

Before concluding this section, I will summarise a few conclusions on "biased" testing. First, if an applied strategy is chosen that increases the chances of the hypothesis being supported, this leads to unwarranted strengthening of prior beliefs, if and only if the tester does not take into account the consequences of this strategy for the evaluation of the supporting evidence (Klayman, 1995). If he does interpret the confirmations with the appropriate scepticism, nothing is wrong. Second, having strong feelings about a hypothesis does not necessarily entail pursuing its confirmation at the cost of the credibility of such a confirmation. In contrast, we even seem willing to take higher risks with our favourite idea, in some contexts. Third, some seemingly "biased" testing behaviour according to formal theories are sometimes defended as *pragmatically* rational. This is a frequently defended stance. This argument points at the possible inadequate application of various formal and logical theories to psychological reality (Evans, 1993; Evans & Over, 1996a; Friedrich, 1993). However, looking more closely, many formal theories of testing explicitly attribute a legitimate role to what are called psychological and pragmatic biases (utilities, prior beliefs). This is especially true for statistical-inference theory (see Chapter 3), thus making the polarisation between pragmatic and formal reasoning somewhat elusive. I shall return to this in the final chapter.

SUMMARY AND CONCLUSIONS

For the psychological study of hypothesis testing in realistic contexts, the statistical approach assuming uncertainties and utilities is a very adequate tool. Indeed, those factors play a role in most everyday information-seeking and hypothesis-testing situations. We seem quite sensitive to the diagnosticity of information, although we attach more importance to the likelihood of data under the hypothesis in which we are interested than to their likelihood if some alternative is true. But if both likelihoods are given, we tend rationally to choose the test that best separates the hypotheses.

With regard to confirmation tendencies, things are much more complicated. Although there is a huge amount of studies claiming that we are generally liable to such a bias when presented with probabilistic and realistic stimuli, the question remains as to what should be understood by this term. If it is a *purposeful*, motivated, frequent, and *successful* attempt to protect our ideas against falsification, then it has not been definitively demonstrated. But less strong variants of the confirmation bias claim have been supported. First, testers prefer extreme likelihoods for the focal hypothesis, and fail to correct for the interpretation of this supporting outcome of these tests *a posteriori*. This has been demonstrated in studies using very abstract and formally defined tests. Thus, it is unclear whether this is done on purpose, and whether this would be replicated in more realistic tasks. Second, in many realistic social situations, in which tests consist of asking questions, the choice of the question triggers a tendency in the respondent to answer positively. This might result in a confirmation bias as well. Third, when available information is to be selected to evaluate a favourite hypothesis, a bias might also occur because only confirming information is believed. However, this behaviour is not "testing" behaviour, but information interpretation. Finally, confirmation tendencies have been equated with considerations of error costs. We might "prefer" to make the error of accepting rather than of rejecting a desirable hypothesis. But this "bias" is perfectly in line with decision theory, as long as we know what we are doing and the consequences of it. There are indications that we seem able to realise and understand the idea of asymmetric error minimisation and the consequences for the value of the proof, when the options are formulated in conceptual terms rather than probabilities. Surprisingly, liking one's own hypothesis very much sometimes even induces a tendency to subject it to more risky tests, because we want to get the strongest possible proof of its truth. This is in line with the probability value model.

On the whole, confirmation tendencies are not straightforward, neither theoretically nor empirically (see also Klayman, 1995). However, the statistical approach and especially recent analyses have pointed towards interesting phenomena occurring in everyday hypothesis testing: (pseudo-) diagnosticity, extremity, test choice and interpretation relations, information bias, probability, and utility-of-outcome considerations. The research programme on hypothesis testing using statistical theories for explaining realistic testing situations seems a promising way to generate theoretically well-grounded, interesting predictions on the hypothesis-testing strategies of everyday reasoners.

CHAPTER SIX

The probability value perspective

RATIONALITY AND BIAS IN TESTING

Are people "rational" or not when they test their ideas against reality? Are they enlightened falsifiers or conservative confirmers? And what is rational? These are the basic issues at stake in most research on human hypothesis testing and, likewise, also gave the impulse to the present study. Throughout this essay, these questions have been treated at different levels. First, the theoretical level. I derived the striking equivalence between a confirming and a falsifying strategy, because whoever falsifies confirms, and whoever confirms, necessarily falsifies. It was demonstrated in the philosophical section that the two standards of testing, although emanating from the rival philosophies of logical positivism and falsificationism, surprisingly prescribe exactly the same testing behaviour.

This conclusion was based on the demonstration that the formal measures of "degree of confirmation" and "degree of severity" of a test are equivalent. Thus, the supposedly incompatible philosophies propose the same normative strategy for the testing of individual hypotheses. This "unified" standard prescribes that one should choose the test where a supporting outcome has the greatest supporting evidential value. This value is the ratio of the extent to which an expected observation is predicted by the hypothesis (its conditional probability) to its probability anyway (its unconditionalised probability). Such a test is most severe for the hypothesis and has the highest corroborating potential, according to falsificationism.

But this very same test also has the highest confirming potential, according to verificationism. And maximising the expected degree of corroboration and the expected degree of confirmation is what these philosophies respectively prescribe. This analysis sheds an interesting light on the possibly most well-known opposition in the philosophy of science and therefore on the psychology of hypothesis testing governed by this theoretical opposition.

Second, the question of whether people are confirming or falsifying testers can be considered from an intuitive point of view. Asserting that people are liable to a confirmation bias implies that they would simultaneously want to seek new information about their hypotheses and want to avoid that information. Imagine a person with frequent headaches. A friend advises him or her to take a certain pill to relieve the symptoms. However, just to be sure, he or she wants to test this advice. Now, testing this hypothesis with a confirmation bias is looking for a test that the person knows will give him or her the positive answer that the pill will work (asking the pharmacist who sells this brand of pill, for example, rather than a doctor). Indeed this bias will probably confirm the person's "favourite" hypothesis, but his or her very intention was to obtain some as yet unknown information about this pill. The example shows that the very intention to perform testing behaviour is somehow at odds with manipulating the test in such a way that only the desired evidence is obtained. This has raised the question as to whether or not confirmation bias is possible at all. I have argued in the present essay that a confirmation bias in testing can best be thought of in terms of the probability value model. I come back to this. In this intuitive line of thought, an evolutionary argument can be added. From the evolutionary angle, a simple confirmation bias is quite implausible. Indeed, if humans would entirely determine top down what reality is and what it is not, they could be prey to their own ideas. Testing the hypothesis that an approaching animal is a dog, by means of a confirmation bias (because we would prefer it to be a dog), whilst it is actually a wolf, would be a fatal strategy.

The third level is the empirical level. Do people actually test their hypotheses in some confirmatory way? The numerous studies discussed in this book do not allow the conclusion that people are motivated conservative testers. Moreover, the previous theoretical arguments against the very idea of wanting confirmations alone were supported in a number of experimental studies discussed in the previous chapters. For example, subjects instructed to apply a falsifying hypothesis-testing strategy felt unable to do so. They reported that they expected confirming test results, even when trying hard to falsify. Furthermore, actually eliminating hypotheses on the basis of counter-evidence did not in itself help testers to discover the truth. From a number of studies, the picture emerges that testers are relatively well served by confirming results when they are not sure about their hypotheses, whereas falsifications are more useful when testers have already acquired some degree of confidence in their hypotheses. Thus, when scrutinised from these three points of view, the "classical" confirmation bias as a characteristic of testing behaviour does not convincingly withstand dissection.

However, a number of claims about pervasive confirmation or verification tendencies originate from the selection task literature. This paradigm forms a special category of hypothesis-testing research based on proposition logic. As expounded in Chapter 4, hundreds of experiments with the selection task demonstrated that people fail to realise the superiority of falsifications for evaluating "if P then Q" statements. They tend to look for "confirming" evidence (cases of P and Q) when they are asked to test the statement "if P then Q", in spite of the low potential of P and Q cases to say something about the conditional statement. What, then, can be concluded from selection task research? First, it was noticed that proposition logic—the "world" in which this standard holds—is a highly artificial formal system, made up of sharply defined operators (connectives) that happen to have some commonalties with natural language, but which are not meant to cover precisely natural language definitions. Interpreting proposition logic connectives as natural language, and applying these formal rules to everyday reasoning, can lead to awkward intuitions. Thus, the proposition logic model for everyday human hypothesis-testing behaviour is not without problems in the first place.

Second, the review of selection task research in Chapter 4 has shown that people are not always non-normative, and that even the apparently incorrect test choices can readily be argued to be normative when analysed from a theoretical perspective other than that which the proposition logical model provides. Some studies try to find and explain facilitation effects, others are more directed towards explaining the incorrect response. In the former, subjects appear to give the "rational" falsifying response, due to some manipulation. The question, then, is, what is the discriminating characteristic of the manipulations that elicit correct testing behaviour? The picture emerges, from this paradigm, that facilitation occurs when the falsifying case (not-Q) somehow catches the tester's attention. Thus, the crucial characteristic for facilitation is that the possible falsifying instances gain relevance in the tester's mind. In that case, the test that can uncover that falsifying case will be chosen. In contrast to what is often assumed, a realistic content is not a sufficient condition for this facilitation, nor a necessary condition. Our attention can be guided towards the falsifications in numerous ways. For example, if the "P-then-Q" hypothesis is framed as a deontic statement, then violations of the rule will come to our mind, and will be subsequently sought. These possible violations receive even more attention when they have a high practical utility for the tester.

Falsifications may also gain relevance when we can represent them mentally more easily than the confirming cases. This happens when the falsifications are the affirmative counterpart of a negated expression. Also, a particular context may point towards the exceptions to some rule. When the tester thinks there is a lot of supporting evidence for the statement, but only a few exceptions, this may motivate the tester to look for these exceptions. Conversely, the exceptions might be found among the confirming Q cases, when the statement is about these very rare cases. Globally, this "set-size effect" is consistent with a recent key analysis of the selection task that maintains that people seek the most "informative" observation to evaluate their hypotheses, independent of its status according to truth tables. Informativeness is inversely proportional to likeliness. Accordingly, the rare cases might be more preferable when they are literally perceived as "exceptional". Thus, the explanations for facilitation in the selection task seem, in general, to converge towards one overarching explanation that is "relevance of the falsification".

In addition to explaining facilitation, a huge number of studies try to explain why the incorrect response (selection of P and Q cards) is generally observed. In the early selection task, failing to select the not-Q item was explained in terms of the shortcomings of human reasoning. Subjects were suspected to have "defective mental truth tables". Recent work, however, tends to consider this response as rational, it being a different kind of rationality than that proposed by Wason and his colleagues. It is statistical rationality rather than logical rationality. Indeed, researchers have turned to "statistical inference" accounts of the selection task. Two major models are discussed in this book. The first assumes that subjects calculate the informativeness of test outcomes and choose the most informative ones. This model indeed predicts the observed non-logical preference for tests that possibly "confirm" the statement when some additional assumptions are satisfied. One of these assumptions is that P and Q cases are perceived by the tester as being relatively rare. This converges with the set-size effect discussed earlier. However, this model is criticised because of the complex calculations it requires and the assumptions it makes.

Another way of modelling selection task performance in statistical terms is to assume that subjects apply the maximum-likelihood principle. This criterion is most commonly used in statistical hypothesis-testing theories. However, this model predicts that the P and not-Q tests should be selected, just as proposition logic did. And this prediction is obviously at odds with the actual response. In sum, the statistical models of hypothesis testing behaviour in the selection task do not provide us with one entirely satisfying account of what people do when faced with this task. The most interesting aspect of these models was argued to be the very fact that the psychology of hypothesis testing is approached from the statistical perspective. This approach represents a promising shift in the investigation into hypothesis testing. Indeed, statistical assumptions like ambiguity, uncertainty, informativeness, probability, and utility are suitable tools to describe an everyday testing situation. But it is questionable whether the selection task is an appropriate task to test this model. Indeed, many parameters of the statistical-inference theory cannot be easily estimated in the selection task situation. For example, how to express the fact that the conditional has "a few exceptions" in probabilities? Proposition logic is designed to handle unambiguous situations, whereas the statistical approaches aim at explaining testing in uncertain and ambiguous situations. Although the last word has not been uttered about the way people fulfil the selection task, hypothesis-testing research can gain substantial benefits from the statistical shift made by its researchers.

The recent interest in alternative though formal models of the selection task, confronts researchers in the field of hypothesis testing again with the question of rationality. Indeed, if the selection task is not performed as the logical standard prescribes, but if this performance satisfies another theoretical background, then what should be concluded about the rationality of hypothesis-testing behaviour? Evans and Over (1996a) propose a view on human rationality that addresses this problem. They distinguish between personal rationality (rationality$_1$) and impersonal rationality (rationality$_2$). Rationality$_1$ refers to cognitive processes that are generally reliable and efficient for achieving one's goal. It is generally a tacit process. Rationality$_2$, in contrast, is a consciously justifiable process sanctioned by a normative theory. In the context of hypothesis testing, and especially the selection task, test selection is supposed to proceed according to rationality$_1$. That is, the tester has some goal, for example, information gain, or finding the rule violators, and selects heuristically according to these goals. However, when the test is performed, the justification people give for their choice is a rationality$_2$ process. But even with the solution Evans and Over propose, the question remains open how rational is behaving according to probabilistic models but not according to logical models (especially in the selection task).

Pragmatic goal-oriented processes and impersonal rationality are possibly less easy to separate in thinking about testing behaviour, than in Evans and Over's dual theory of rationality. For example, a tester can have the pragmatic goal of getting as much information as possible at the lowest cost, while nonetheless choosing a test that is not optimal in this respect according to some formal theory. Should we conclude that the tester is pragmatically rational or not? In other words, people may have goals that are demonstrably best reached by some formal method, but still do not choose this method best fitting their own goal. The goal being pragmatic does not guarantee that reasoning will be efficient with regard to this goal.

Actually, this problem is ubiquitous in much pragmatic reasoning, in which people aim to be accurate, but fail to be because they "neglect base rates", they generate too few alternative hypotheses, they do not consider the alternative likelihood, or they overestimate the value of a result, and so on. A very nice example is the effect Slowiaczek et al. demonstrated: People who choose low proof-value tests for their hypotheses, overestimate the value of the proof once it is there.

In summary, Evans and Over's proposal to distinguish two forms of rationality is the best worked out theory of rationality in reasoning, at the moment, although the contrast between the two kinds of rationality is somewhat challenged by the statistical inference turn in hypothesis testing, and reasoning research. Indeed, statistical models are essentially pragmatic, because of the slots in the formulas left open for prior beliefs and error utilities, but still are theoretically perfectly coherent. Therefore, this new approach has essentially diluted the contrast between pragmatic considerations and strictly normative considerations.

THE PROBABILITY VALUE MODEL

If the falsification–confirmation distinction is not wholly tenable, and rational testing not easily distinguishable from non-rational testing, how should we proceed to study hypothesis testing? On which dimensions should it be observed? A psychological model is derivable from the theories of testing, which is more sophisticated than the old confirmation–falsification distinction. This is the "probability value" model, introduced in Chapter 5. Indeed, in the present study, most theories of hypothesis testing have been shown to have an essential commonality. They basically describe a test with regard to the probabilities of its possible outcomes and the values of these outcomes to support or reject the focal hypothesis. Under general conditions, the probability of an outcome is inversely proportional to its value in supporting or rejecting a hypothesis.

The probability value model is the psychological counterpart of the unified principle of severity and confirmation (Chapter 1). This principle underlies almost all testing theories. This of hypothesis testing is an attempt to integrate a number of theories and findings discussed in this essay, and to solve controversies and paradoxes like the confirmation-bias paradox (Poletiek, 1996). Although I believe it to be a promising perspective on hypothesis testing behaviour for which some support has already been found (Poletiek, 1995; Poletiek & Berndsen, 2000), it still needs further testing in future research. In the following, I describe the contours of this view on testing behaviour, and a few implications. Hypothesis testing, defined in the introduction as looking for new information about some idea, might be represented as solving two dilemmas. The first one is, "what kind of evidence can I get at what cost?", and the second one is, "how many decision errors of one kind do I want to avoid at the cost of how many decision errors of the other kind?". These dilemmas may be solved in either order. A tester might be concerned first with the overall costs of searching for evidence, and next with the costs of a particular kind of error with regard to his or her hypothesis about the state of nature. But it may also be the other way round: He or she might be prepared to pay whatever costs to avoid a certain error in his ideas about the world, no matter what this may imply for the chance of making the other type of error. These dilemmas can be recognised in realistic test situations, as well as the task situations discussed in the previous chapters. This representation of the task of a hypothesis tester is "evolutionary" in the sense that the tester is seen as dealing with evidence from the external world, to which he or she has to submit his ideas, but always with some ultimate goal (Evans & Over, 1996a). This goal controls what and how much contribution will be accorded to the external evidence and how much is left for internal belief in this interaction.

The first dilemma might also be called the symmetrical dilemma and the second the asymmetrical dilemma of testing. Let us assume a simple testing situation with a hypothesis H, a test with two possible outcomes, one supporting the hypothesis (c) and the other disconfirming (d). The first dilemma might be described formally by means of its expected subjective utility as follows:

$$\mathrm{SEUt} = \mathrm{p(c)\ v(c) + p(d)\ v(d) - costs} \tag{6.1}$$

in which the subjective expected utility (SEU) of a test is a function of its power to confirm *and* disconfirm the hypothesis, the outcome probabilities, and the costs of performing it. The terms p and v should be seen as subjective estimations of these probabilities and values. The dilemma consists in making a trade-off between costs and the information expected from testing. This is how Klauer (1999) represents sequential testing behaviour in the selection task. In social hypothesis-testing research, this dilemma corresponds to accuracy of tests (Trope & Liberman, 1996). Without costs for testing there is hardly any dilemma here. However, these costs can be psychological, financial, or of any kind. Also, this description of the test choice dilemma is not meant to be normative.

The second, asymmetrical dilemma can be described as follows. Consider a number of tests with equal expected utility and equal costs. We also assume an inverse relation between values and probabilities of outcomes of a test (Howson & Urbach, 1989). The tester facing the dilemma to choose among those tests can either maximise the probability of one outcome at the cost of its value or vice versa. The values of the outcomes express the probabilities of decision errors tolerated about the hypothesis. Thus, the dilemma consists in finding a trade-off between the probability of getting some kind of result from a test and its value to (dis)confirm the hypothesis, as argued in the previous chapter.

The probability value model of testing refers mainly to the second dilemma. We might represent it in terms of game theory. A game can be described as having outcomes with some values and probabilities. Suppose, for example, that the player decides to pay 15 euros for a game from which he can expect to win 10 euros (the first trade off); this represents the player's subjective utility of the game. The second dilemma is what kind game he wants to play. What kind of risks does he or she want to take: a high risk to win few, a low risk to win much, or average risks to win average amounts? The following two games, which both provide an expected utility to win 10 euros, describe this choice.

Game 1	1% (900€) + 99% (1€)
Game 2	25% (24€) + 75% (6€)

The player preferring the first game is called a risk seeker. The one preferring the second game is more risk averse. Analogously, a tester can either go for a test providing highly probative evidence but with a high risk of finding only weakly rejecting evidence, or a test with a reasonable probability of finding one of both results. This is how Poletiek and Berndsen (2000) analysed the test choice of their participants in a hypothesis-testing task with realistic materials. They found that the testers' choice depends on their belief and commitment regarding the hypothesis. Interestingly, favouring one hypothesis did not make participants maximise the probability of getting the confirming outcome, which was the idea behind confirmation bias. In contrast, testers were sometimes risky evidence seekers when the hypothesis was believed to be true.

I believe that the two dilemmas refer to psychologically different processes. The question of what we are prepared to "pay" for whatever new information about a hypothesis, is not the same as what kind of information we look for. The first dilemma might be studied by observing how many tests people perform, when each test has a price (Poletiek, 1996; Van Wallendael, 1995). The second one refers to which tests they propose. Also, the symmetrical and asymmetrical models of the selection task assume a different kind of test choice dilemma (Klauer, 1999). The second aspect of testing behaviour is more psychological; the first one almost economic. Also it represents the tester as having no interests in one or the other outcome. The asymmetrical models try to deal with the question which lies at the core of most hypothesis-testing research: confirmation tendencies, and sensitivity to disconfirmation. This probability value model of testing is quite intuitive. Indeed, it reveals that we can either choose to be "risk averse" with regard to a theory, testing it conservatively against some evidence that we are "reasonably sure" will provide an affirmative result, but which also gives little more "confidence" in the theory. Or we might be "risk seeking", going for a "convincing" empirical proof that we know, in advance, has little chance of showing up. The information we seek is either highly probable and poorly convincing, or surprising and highly convincing. Also, in real life we feel that there are some false theories we can live with, without suffering substantial damages; for others we want to be sure they correspond with the state of nature.

Modelling testing behaviour in this way solves the intuitive paradox we pointed out when discussing the standard "confirmation bias" findings. Indeed, imagine again the patient with the pill. He or she might want to maximise the probability of a positive answer, taking the low probative significance of such an answer for granted. The patient's attitude with regard to the pill is "better something than nothing against this headache, even if it is not fully cured". He or she goes for the low risk strategy. But, by doing so, the patient simultaneously maximises the value of a negative answer, being a falsifier in this sense. Indeed, the patient can be very confident about the pill's ineffectiveness with regard to the headache if the pharmacist, quite surprisingly, gives him or her a negative answer—the pill will not help. Defined in this way, a "confirmation bias" is no longer illogical or intuitively awkward. This is because, in this definition, the tester who maximises the likeliness of hearing the desired answer also presumably takes the implications for granted that such an answer tells little.

A remarkable implication of the probability value model was pointed out in Chapter 2. It regards the Popperian paradox. Popper argued for the search of falsifications by means of severe testing. The method of falsifying theories was ultimately the best way to make knowledge grow: Every rejection was a positive step in science, and should be seen by the scientist as a happy result. But it follows from the present analysis that a test that maximises the probability of a certain result generally minimises its value when one actually procures it. The zealous Popperian scientist who obtains the pursued counter-evidence against his or her hypothesis will, probably, be highly disappointed when he or she finds it. It will tell him or her little. Thus, looking for falsifications is nice as long as one does not get them. Popper's precept to try to falsify, and to be happy when one succeeds, is basically contradictory. In this context, I demonstrated that the probability value model can also be applied in the rule discovery task, by observing how risky test choices are. The results suggested that being very risky can generate too many falsifications to be helpful. However, performing moderately risky tests, especially at the onset of the discovery process, seems more efficient for discovery performance. In the selection task, however, the probability value model is not easily applicable, for the same reasons as why other statistical models were not unproblematic. The probability value perspective fits in a more realistic view on hypothesis testing, unconstrained by logical laws.

A nice consequence of the probability value model is that we can identify the cognitive factor from the "objective" factor in hypothesis-testing behaviour, but also that both are dependent on each other. The impact of nature after testing will be bigger as the ambiguity of the observations is smaller, but also as the tester decides to gather more evidence. Conversely, the influence of human strategy is larger as the tester has stronger feelings about the risks and values he or she demands and tolerates with regard to the test results. These considerations can lead to accepting a hypothesis not fully confirmed, but about which false beliefs do not represent any danger. From the probability value perspective, traditional hypothesis-testing research can be said to focus merely on the probability side. Indeed, past research on confirmation biases showed that testers bias their tests in such a way as to find support. But this research could not account for the relative impact of nature versus the human, in our interaction with the world. Indeed, if we really tried to minimise the probability of falsifications anyway, we would not survive nature, ultimately. Clearly, recent approaches do take into account the value of evidence judged according to concepts such as "epistemic utility", "maximising information gain", and "answer diagnosticity", which are emerging in the field. In sum, the probability value perspective might be a promising guideline for future hypothesis-testing research.

APPLICATIONS AND FUTURE RESEARCH

Hypothesis-testing research has been carried out mostly with abstract tasks. But what is it worth in real-life situations? Spontaneous, everyday hypothesis testing is difficult to study in a natural setting: Humans reason tacitly and even implicitly. A step towards more ecological validity would be to carry out experiments simulating everyday hypothesis-testing situations or, perhaps, quasi-experimental studies in realistic settings. Observations could then be made of how people solve the probability value dilemma under different conditions. These conditions should reflect many aspects of everyday testing situations. For example, how important is outcome reliability to a tester? What is the tester's position with regard to rejection and support, and does this position change after the evidence has been observed? Under which conditions is the value of a piece of evidence maximised, and under which conditions the probability? Do testers revise their belief in the hypothesis and act in accordance with the evidential value of the outcome? Personal characteristics may also influence the way people solve the probability value dilemma. Which kinds of people are conservative (rejection-risk aversive) in evaluating their beliefs and which kinds go for counter-evidence (rejection-risk seeking)?

Hypothesis testing is at the core of much expert reasoning. In many professional situations, hypotheses have to be tested systematically. The most frequently studied professional situation is probably that involving medical diagnosis. Clearly, physicians constantly test hypotheses about the causes of their patients' symptoms. For example, one conclusion of this field of research, which might be related to the present analysis, is that experts are often overconfident in their faith in the diagnosis after having examined the patient (e.g., Christensen-Szalansky & Bushyhead, 1981). This overconfidence has been attributed to confirmation bias (Koriat, Lichtenstein, & Fischhoff, 1980). This explanation means that people first search in their mind for relevant knowledge when they have to evaluate the truth of a hypothesis. Subsequently, they decide about the truth status of the hypothesis, and finally, they assess their confidence in this decision. In this view, a kind of mental hypothesis testing has taken place. Our analysis suggests that the overestimation of the value of the (mentally retrieved) test results may cause the overconfidence. Experts, like doctors, might overestimate the value of evidence observed or retrieved from the mentally performed tests, taking insufficient account of the properties of the "test" that generated the data in the first place.

Another authority even more explicitly involved in hypothesis testing is the legal expert. The crucial task of a judge (or a jury) is to test the hypothesis concerning the guilt of a defendant. Performing this test in an adequate and fair way, giving the evidence the weight it deserves in order to evaluate the hypothesis, is a highly pertinent case of professional hypothesis testing. Interestingly, the value of the evidence must be explicitly accounted for in legal decision making. Dutch law requires "legal and convincing evidence" to support a decision in court. According to English law, a jury should test whether the value of the evidence is such that it can prove the guilt hypothesis to be true "beyond reasonable doubt". Thus, correctly estimating and justifying the value of evidence is essential in jurisdiction. In terms of the probability value model of testing, a judge or jury should take into account the test properties from which the evidence brought into court has issued. In a study by Wagenaar, Van Koppen, and Crombag (1993), striking examples of biased hypothesis testing in legal decision making are presented. For example, when the interrogation of witnesses is chosen as the test of the hypothesis concerning guilt, some answers of these witnesses might very likely simply be due to the interrogation method, or due to the witness's sensitivity to pressure, feeling that a particular answer is desirable. Such a test generates low value confirmations. This does not necessarily lead to biased testing as long as the judge is aware of the consequences of these test characteristics for the low evidential value of the results, and takes these consequences into consideration in the final judgement. Clearly, the legal hypothesis-testing situation lends itself very well to a probability value analysis.

Hence, the probability value perspective might be one of the challenges of future hypothesis-testing research, paying rich dividends in both theory and actual practice.

ACKNOWLEDGEMENT

This publication has been supported financially by The Netherlands Organisation for Scientific Research (NWO).

References

Almor, A., & Sloman, S.A. (1996). Is deontic reasoning special? *Psychological Review*, *103*, 374–380.

Baron, J. (1985). *Rationality and intelligence*. Cambridge: Cambridge University Press.

Baron, J., Beattie, J., & Hershey, J.C. (1988). Heuristics and biases in diagnostic reasoning: II. Congruence, information, and certainty. *Organizational Behavior and Human Decision Processes*, *42*, 88–110.

Bassok, M., & Trope, Y. (1984). People's strategies for testing hypotheses about another's personality: Confirmatory or diagnostic? *Social Cognition*, *2*, 199–216.

Beattie, J., & Baron, J. (1988). Confirmation and matching biases in hypothesis testing. *Quarterly Journal of Experimental Psychology*, *40A*, 269–297.

Bechtel, W. (1988). *Philosophy of science: An overview for cognitive science*. Hillsdale, NJ: Lawrence Erlbaum Associates Inc.

Beyth-Marom, R., & Fischhoff, B. (1983). Diagnosticity and pseudodiagnosticity. *Journal of Personality and Social Psychology*, *45*, 1185–1195.

Braine, M.D.S. (1978). On the relation between the natural logic of reasoning and standard logic. *Psychological Review*, *85*, 1–21.

Bruner, J.S., Goodnow, J.A., & Austin, G.A. (1956). *A study of thinking*. New York: Wiley.

Carnap, R. (1928). *Der Logische Aufbau der Welt.* Berlin: Weltkreis-Verlag.

Carnap, R. (1936–37). Testability and meaning. *Philosophy of Science*, *3*, 420–468; *4*, 1–40.

Carnap, R. (1950). *Logical foundations of probability*. Chicago, IL: University of Chicago Press.

Chater, N., & Oaksford, M. (1999) Information gain and decision-theoretic approaches to data selection: Response to Klauer (1999). *Psychological Review*, *106*, 223–227.

Cheng, P.W., & Holyoak, K.J. (1985). Pragmatic reasoning schemas. *Cognitive Psychology*, *17*, 391–416.

Cheng, P.W., & Holyoak, K.J. (1989). On the natural selection of reasoning theories. *Cognition*, *33*, 285–313.

Christensen-Szalansky, J.J.J., & Bushyhead, J.B. (1981). Physicians' use of probabilistic information in a real clinical setting. *Journal of Experimental Psychology: Human Perception and Performance*, *7*, 928–935.

Coombs, C.H., Daws, R.M., & Tversky, A. (1970). *Mathematical psychology*. Englewood Cliffs, NJ: Prentice Hall.

Cosmides, L. (1989). The logic of social exchange: Has natural selection shaped how humans reason? Studies with the Wason selection task. *Cognition*, *31*, 187–276.

De Groot, A.D. (1969). *Methodology: Foundations of inference and research in the behavioral sciences*. The Hague: Mouton.

Doherty, M.E., & Mynatt, C.R. (1990). Inattention to P (H) and to P (D | ~H): A converging operation. *Acta Psychologica*, *75*, 1–11.

Doherty, M.E., Mynatt, C.R., Tweney, R.D., & Schiavo, M.D. (1979). Pseudodiagnosticity. *Acta Psychologica*, *43*, 111–121.

Erwin, E., & Siegel, H. (1989). Is confirmation differential? *British Journal for the Philosophy of Science*, *40*, 105–119.

Evans, J.St.B.T. (1972). Interpretation and matching bias in a reasoning task. *Quarterly Journal of Experimental Psychology*, *24*, 193–199.

Evans, J.St.B.T. (1982). *The psychology of deductive reasoning*. London: Routledge & Kegan Paul.

Evans, J.St.B.T. (1983). Linguistic determinants of bias in conditional reasoning. *Quarterly Journal of Experimental Psychology*, *35A*, 635–644.

Evans, J.St.B.T. (1984). Heuristic and analytic processes in reasoning. *British Journal of Psychology*, *75*, 451–468.

Evans, J.St.B.T. (1989). *Bias in human reasoning*. Hove, UK: Lawrence Erlbaum Associates Ltd.

Evans, J.St.B.T. (1991). Theories of human reasoning: The fragmented state of the art. *Theory and Psychology*, *1*, 83–105.

Evans, J.St.B.T. (1993). The cognitive psychology of reasoning: An introduction. [Special Issue.] *Quarterly Journal of Experimental Psychology Human Experimental Psychology*, *46A*, 561–567.

Evans, J.St.B.T. (1994). Relevance and reasoning. In S. Newstead & J.St.B.T. Evans (Eds.), *Current directions in thinking and reasoning*. Hillsdale, NJ: Lawrence Erlbaum Associates Inc.

Evans, J.St.B.T. (1998) Matching bias in conditional reasoning: Do we understand it after 25 years? *Thinking and Reasoning*, *4*, 45–82.

Evans, J.St.B.T., & Lynch, J.S. (1973). Matching bias in the selection task. *British Journal of Psychology*, *64*, 391–397.

Evans, J.St.B.T., & Over, D.E. (1996a). *Rationality and reasoning*. Hove, UK: Psychology Press.

Evans, J.St.B.T., & Over, D.E. (1996b). Rationality in the selection task: Epistemic utility versus uncertainty reduction. *Psychological Review*, *103*, 356–363.

Evans, J.St.B.T., Newstead, S.E., & Byrne, R.M. (1993). *Human reasoning: The psychology of deduction*. Hove, UK: Lawrence Erlbaum Associates Ltd.

Farris, H.H., & Revlin, R. (1989). Sensible reasoning in two tasks: Rule discovery and hypothesis evaluation. *Memory and Cognition*, *17*, 221–232.

Feyerabend, P. (1970). Against method: Outline of an anarchistic theory of knowledge. In M. Radner & S. Winokur (Eds.), *Minnesota studies in the philosophy of science* (pp. 17–130). Minneapolis, MN: University of Minnesota Press.

Fiedler, K., & Hertel, G. (1994). Content-related schemata versus verbal-framing effects in deductive reasoning. *Social Cognition*, *12*, 129–147.

Fischhoff, B., & Beyth-Marom, R. (1983). Hypothesis evaluation from a Bayesian perspective. *Psychological Review*, *90*, 239–260.

Fisher, R. (1950). *Statistical methods for research workers* (11th ed.). Edinburgh: Oliver and Boyd.

Fisher, R. (1959). *Statistical methods and scientific inference* (2nd ed.). London: Oliver and Boyd.

Foss, B.M. (Ed.). (1966). *New horizons in psychology: I.* Harmondsworth, UK: Penguin.

Friedrich, J. (1993). Primary error detection and minimization (PEDMIN) strategies in social cognition: A reinterpretation of confirmation bias phenomena. *Psychological Review*, *100*, 298–319.

Gadenne, V. (1982). Der Bestätigungsfehler und die Rationalität kognitiver Prozesse. *Psychologische Beiträge*, *24*, 11–25.

Gadenne, V., & Oswald, M. (1986). Entstehung und Veränderung von Bestätigungstendenzen beim Testen von Hypothesen [Formation and alteration of confimatory tendencies during the testing of hypotheses]. *Zeitschrift für Experimentelle and Angewanste Psychologie*, *33*, 360–374.

Gamut, L.T.F. (1991). *Introduction to logic*. Chicago, IL: Chicago University Press.

Giere, R.N. (1975). The epistemological roots of scientific knowledge. In R.M. Anderson, Jr. & G. Maxwell (Eds.), *Induction, probability and confirmation. Minnesota studies in the philosophy of science* (pp. 212–261). Minneapolis, MN: University of Minnesota Press.

Giere, R.N. (1977). Testing vs information models of statistical inference. In R.G. Colodny (Ed.), *Logic, laws and life: Pittsburg series in the philosophy of science, Vol. 6* (pp. 19–70). Pittsburgh, PA: University of Pittsburgh Press.

Giere, R.N. (1988). *Explaining science: A cognitive approach*. Chicago, IL: University of Chicago Press.

Gigerenzer, G., & Hug, K. (1992). Domain-specific reasoning: Social contracts, cheating, and perspective change. *Cognition*, *43*, 127–171.

Gigerenzer, G., & Murray, D.J. (1987). *Cognition as intuitive statistics*. Hillsdale, NJ: Lawrence Erlbaum Associates Inc.

Golding, E. (1981). *The effect of past experience on problem solving*. Paper presented to the British Psychological Society at Surrey University.

Gorman, M.E. (1989). Error, falsification and scientific inference: An experimental investigation. *Quarterly Journal of Experimental Psychology*, *41A*, 385–412.

Gorman, M.E. (1995). Confirmation, disconfirmation, and invention: The case of Alexander Graham Bell and the telephone. *Thinking and Reasoning*, *1*, 31–53.

Gorman, M.E., & Gorman, M.E. (1984). A comparison of disconfirmatory, confirmatory and control strategies on Wason's 2–4–6 task. *Quarterly Journal of Experimental Psychology*, *36A*, 629–648.

Gorman, M.E., Gorman, M.E., Latta, R.M., & Cunningham, G. (1984). How disconfirmatory, confirmatory and combined strategies affect group problem solving. *British Journal of Psychology*, *75*, 65–79.

Gorman, M.E., Stafford, A., & Gorman, M.E. (1987). Disconfirmation and dual hypotheses on a more difficult version of Wason's 2–4–6 task. *Quarterly Journal of Experimental Psychology*, *39*, 1–28.

Green, D.W. (1990). Confirmation bias, problem-solving and cognitive models. In C. Jean-Paul, F. Jean-Marc, & G. Michel (Eds.), *Cognitive biases* (pp. 553–562). Amsterdam: North-Holland.

Green, D.W., & Over, D.E. (1997). Causal inferences, contingency tables and the selection task. *Cahiers de Psychologie Cognitive*, *16*, 459–487.

Green, D.W., Over, D.E., & Pyne, R.A. (1997) Probability and choice in the selection task. *Thinking and Reasoning*, *3*, 209–235.

Griggs, R.A., & Cox, J.R. (1982). The elusive thematic-materials effect in Wason's selection task. *British Journal of Psychology*, *73*, 407–420.

Halberstadt, N., & Kareev, Y. (1995). Transitions between modes of inquiry in a rule discovery task. *Quarterly Journal of Experimental Psychology Psychology*, *48A*, 280–295.

Hanson, N.R. (1958). *Patterns of discovery*. Cambridge: Cambridge University Press.

Hanson, N.R. (1969). *Perception and discovery*. San Francisco: Freeman, Cooper and Co.

Hardman, D. (1998). Does reasoning occur on the selection task? A comparison of relevance-based theories. *Thinking and Reasoning*, *4*, 353–376.

Hays, W.L. (1973). *Statistics for the social sciences* (2nd ed.). New York: Holt, Rinehart and Winston.

Hoch, S.J., & Tschirgi, J.E. (1985). Logical knowledge and cue redundancy in deductive reasoning. *Memory and Cognition*, *13*, 453–462.

Hoel, P.G. (1984). *Introduction to mathematical statistics* (5th ed.) New York: Wiley.

Hofstee, W.K.B. (1980). *De empirische discussie* [The empirical discussion]. Meppel: Boom.

Howson, C., & Urbach, P. (1989). *Scientific reasoning*. Chicago: Open Court.

Jackson, S.L., & Griggs, R.A. (1990). The elusive pragmatic reasoning schema's effect. *Quarterly Journal of Experimental Psychology*, *42A*, 637–652.

Johnson-Laird, P.N. (1983). *Mental models: Towards a cognitive science of language, inference and consciousness*. Cambridge: Cambridge University Press.

Johnson-Laird, P.N., & Byrne, R.M.J. (1991). *Deduction*. Hove, UK: Lawrence Erlbaum Associates Ltd.

Johnson-Laird, P.N., & Byrne, R.M.J. (1992). Modal reasoning, models and Manktelow and Over. *Cognition*, *43*, 173–182.

Johnson-Laird, P.N., Legrenzi, P., & Legrenzi, M.S. (1972). Reasoning and a sense of reality. *British Journal of Psychology*, *63*, 395–400.

Johnson-Laird, P.N., & Wason, P.C. (1970). A theoretical analysis of insight into a reasoning task. *Cognitive Psychology*, *1*, 134–148.

Kahneman, D., Slovic, P., & Tversky, A. (Eds.). (1982). *Judgement under uncertainty: Heuristics and biases*. Cambridge: Cambridge University Press.

Kareev, Y., & Halberstadt, N. (1993). Evaluating negative tests and refutations in a rule discovery task. [Special Issue.] *Quarterly Journal of Experimental Psychology*, *46A*, 715–727.

Kareev, Y., Halberstadt, N., & Shafir, D. (1993). Improving performance and increasing the use of non-positive testing in a rule-discovery task. [Special Issue: The cognitive psychology of reasoning.] *Quarterly Journal of Experimental Psychology*, *46A*, 729–742.

Kelly, G.A. (1955). *A theory of personality: The psychology of personal constructs*. New York: Norton.

Kirby, K.N. (1994). Probabilities and utilities of fictional outcomes in Wason's four-card selection task. *Cognition*, *51*, 1–28.

Klauer, K.C. (1999). On the normative justification for information gain in Wason's selection task. *Psychological Review*, *106*, 215–222.

Klayman, J. (1995). Variaties of confirmation bias. In J.R. Busemeyer, R. Hastie, & D.L. Medin (Eds.), *Decision making from a cognitive perspective* (pp. 385–414). San Diego, CA: Academic Press Inc.

Klayman, J., & Ha, Y. (1987). Confirmation, disconfirmation, and information in hypothesis testing. *Psychological Review*, *94*, 211–228.

Klayman, J., & Ha, Y. (1989). Hypothesis testing in rule discovery: Strategy, structure, and content. *Journal of Experimental Psychology: Learning, Memory, and Cognition*, *15*, 596–604.

Koehler, J.J. (1993). The influence of prior beliefs on scientific judgements of evidence quality. *Organizational Behavior and Human Decision Processes*, *56*, 28–55.

Koriat, A., Lichtenstein, S., & Fischhoff, B. (1980). Reasons for confidence. *Journal of Experimental Psychology: Human Learning and Memory*, *6*, 107–118.

Krueger, L.E., Daston, L.J.E., & Heidelberger, M.E. (1987). *The probabilistic revolution. Vol. 1: Ideas in history*. Cambridge, MA: MIT Press.

Krueger, L.E., Gigerenzer, G., & Morgan, M.S. (1987). *The probabilistic revolution. Vol. 2: Ideas in the sciences*. Cambridge, MA: MIT Press.

Kruglanski, A.W. (1996). Motivated social-cognition: Principles of the interface. In E.T. Higgins & A.W. Kruglanski (Eds.), *Social psychology: Handbook of basic principles*. New York: Guilford.

Kuhn, T.S. (1970). Logic of discovery or psychology of research? In I. Laktos & A. Musgrave (Eds.), *Criticism and the growth of knowledge* (pp. 1–25). London: Cambridge University Press.

Kuhn, T.S. (1973). *The structure of scientific revolutions* (2nd ed.). Chicago, IL: University of Chicago Press.

Kullback, S. (1959). *Information theory and statistics*. New York: Wiley.

Lakatos, I. (1970). Falsification and the methodology of scientific research programmes. In I. Lakatos & A. Musgrave (Eds.), *Criticism and the growth of knowledge* (pp. 91–196). London: Cambridge University Press.

Laming, D. (1996). On the analysis of irrational data selection: A critique of Oaksford and Chater (1994). *Psychological Review*, *103*, 364–373.

Laudan, L. (1977). *Progress and its problems*. Berkeley, CA: University of California Press.

Lindley, D.V. (1965). *Introduction to probability and statistics from a Bayesian viewpoint*. Cambridge: Cambridge University Press.

Lord, C.G., Ross, L., & Lepper, M.R. (1979). Biased assimilation and attitude polarization: The effects of prior theories on subsequently considered evidence. *Journal of Personality and Social Psychology*, *37*, 2098–2109.

Love, R.E., & Kessler, C.L. (1995). Focusing in Wason's selection task: Content and instruction effects. *Thinking and Reasoning*, *1*, 153–182.

Mahoney, M.J. (Ed.). (1976). *Scientists as subject: The psychological imperative*. Cambridge, MA: Ballinger.

Manktelow, K.I. (1981). Recent developments in research on Wason's selection task. *Current Psychological Reviews*, *1*, 257–268.

Manktelow, K.I. (1999). *Reasoning and thinking*. Hove, UK: Psychology Press.

Manktelow, K.I., & Evans, J.S. (1979). Facilitation of reasoning by realism: Effect or non-effect? *British Journal of Psychology*, *70*, 477–488.

Manktelow, K.I., & Over, D.E. (1990). *Inference and understanding: A philosophical and psychological perspective*. London: Routledge.

Manktelow, K.I., & Over, D.E. (1992). Utility and deontic reasoning: Some comments on Johnson-Laird and Byrne. *Cognition*, *43*, 183–188.

Manktelow, K.I., & Over, D.E. (1995). Deontic reasoning. In J.St.B.T. Evans & S.E. Newstead (Eds.), *Perspectives on thinking and reasoning: Essays in honour of Peter Wason* (pp. 91–114). Hove, UK: Lawrence Erlbaum Associates Ltd.

Manktelow, K.I., Sutherland, E.J., & Over, D.E. (1995). Probabilistic factors in deontic reasoning. *Thinking and Reasoning*, *1*, 201–220.

McDonald, J. (1990). Some situational determinants of hypothesis-testing strategies. *Journal of Experimental Social Psychology*, *26*, 255–274.

McDonald, J. (1992). Is strong inference superior to simple inference? *Syntheses*, *92*, 261–282.

Montague, R. (1974). Pragmatics. In R. Thomason (Ed.), *Formal philosophy*. New Haven, CT: Yale University Press.

Murphy, G.L., & Medin, D.L.L. (1985). The role of theories in conceptual coherence. *Psychological Review*, *92*, 289–316.

Mynatt, C.R., Doherty, M.E., & Dragan, W. (1993). Information relevance, working memory and the consideration of alternatives. *Quarterly Journal of Experimental Psychology, 46A*, 759–778.

Mynatt, C.R., Doherty, M.E., & Sullivan, J.A. (1991). Data selection in a minimal hypothesis testing task. *Acta Psychologica, 76*, 293–305.

Mynatt, C.R., Doherty, M.E., & Tweney, R.D. (1977). Confirmation bias in a simulated research environment: An experimental study of scientific inference. *Quarterly Journal of Experimental Psychology, 29*, 85–95.

Mynatt, C.R., Doherty, M.E., & Tweney, R.D. (1978). Consequences of confirmation and disconfirmation in a simulated research environment. *Quarterly Journal of Experimental Psychology, 30*, 395–406.

Newstead, S.E., & Evans, J.St.B.T. (Eds.). (1995). *Perspectives on thinking and reasoning: Essays in honour of Peter Wason.* Hove, UK: Lawrence Erlbaum Associates Ltd.

Nickerson, R.S. (1996). Hempel's paradox and Wason's selection task: Logical and psychological puzzles of confirmation. *Thinking and Reasoning, 2*, 1–31.

O'Brien, D.P. (1993). Mental logic and irrationality: We can put a man on the moon, so why can't we solve those logical reasoning problems? In K.I. Manktelow & D.E. Over (Eds.), *Rationality*. London: Routledge.

Oaksford, M., & Chater, N. (1994a). Another look at eliminative behaviour in a conceptual task. *European Journal of Cognitive Psychology, 6*, 149–169.

Oaksford, M., & Chater, N. (1994b). A rational analysis of the selection task as optimal data selection. *Psychological Review, 101*, 608–631.

Oaksford, M., & Chater, N. (1996). Rational explanation of the selection task. *Psychological Review, 103*, 381–391.

Oaksford, M., Chater, N., & Grainger, B. (1999) Probabilistic effects in data selection. *Thinking and Reasoning, 5*, 193–243.

Oaksford, M., Chater, N., Grainger, B., & Larkin, J. (1997). Optimal data selection in the reduced array selection task. *Journal of Experimental Psychology: Learning, Memory and Cognition, 23*, 441–458.

Oaksford, M., & Stenning, K. (1992). Reasoning with conditional containing negated constituents. *Journal of Experimental Psychology: Learning, Memory and Cognition, 18*, 835–854.

Oberauer, K., Wilhelm, O., & Rosas Diaz, R. (1999). Bayesian rationality for the Wason selection task? A test of optimal data selection theory. *Thinking and Reasoning, 5*, 115–144.

O'Brien, D.P. (1995). Finding logic in human reasoning means looking in the right places. In S.E. Newstead & J.St.B.T. Evans (Eds.), *Perspectives on thinking and reasoning: Essay in honour of Peter Wason.* Hove, UK: Lawrence Erlbaum Associates Ltd.

Ormerod, T.C., Manktelow, K.I., & Jones, G.V. (1993). Reasoning with three types of conditional: Biases and mental models. [Special Issue: The cognitive psychology of reasoning.] *Quarterly Journal of Experimental Psychology, 46A*, 653–677.

Over, D.E., & Evans, J.S. (1994). Hits and misses: Kirby on the selection task. *Cognition, 52*, 235–243.

Phillips, L.D. (1973). *Bayesian statistics for social scientists.* London: Nelson.

Platt, R.D., & Griggs, R.A. (1993). Facilitation in the abstract selection task: The effects of attentional and instructional factors. [Special Issue: The cognitive psychology of reasoning.] *Quarterly Journal of Experimental Psychology, 46A*, 591–613.

Poletiek, F.H. (1995). Testing in a rule discovery task: Strategies of test choice and test result interpretation. In J.P. Caverni, M. Bar-Hillel, H. Barron, & H. Jungermann (Eds.), *Contributions to decision research—I* (pp. 335–350). Amsterdam: North-Holland.

Poletiek, F.H. (1996). Paradoxes of falsification. *Quarterly Journal of Experimental Psychology, 49A*, 447–462.

Poletiek, F.H., & Berndsen, M. (2000). Hypothesis testing as risk taking with regard to beliefs. *Journal of Behavioral Decision Making, 13*, 107–123.

Pollard, P. (1981). The effect of thematic content on the "Wason selection task". *Current Psychological Research, 1*, 21–29.

Pollard, P., & Evans, J.S. (1981). The effects of prior beliefs in reasoning: An associational interpretation. *British Journal of Psychology, 72*, 73–81.

Pollard, P., & Evans, J.S. (1983). The effect of experimentally contrived experience on reasoning performance. *Psychological Research, 45*, 287–301.

Popper, K.R. (1935). *Logik der Forschung*. Wien: J. Springer Verlag.

Popper, K.R. (1959/1974). *The logic of scientific discovery* (3rd ed.) London: Hutchinson.

Popper, K.R. (1963/1978). *Conjectures and refutations* (4th ed.). London: Routledge and Kegan Paul.

Reich, S.S., & Ruth, P. (1982). Wason's selection task: Verification, falsification and matching. *British Journal of Psychology, 73*, 395–405.

Rips, L.J. (1983). Cognitive processes in propositional reasoning. *Psychological Review, 90*, 38–71.

Rumelhart, D.E. (1980). Schemata: The building blocks of cognition. In R.J. Spiro, B.C. Bruce, & W.F. Brewer (Eds.), *Theoretical issues in reading comprehension*. Hillsdale, NJ: Lawrence Erlbaum Associates Inc.

Salmon, W.C. (1973). Confirmation. *Scientific American, 228*, 75–83.

Salmon, W.C. (1975). Confirmation and relevance. In G. Maxwell & R.M. Anderson (Eds.), *Minnesota studies in the philosophy of science, VI* (pp. 3–36). Minneapolis, MN: University of Minnesota Press.

Salmon, W.C. (1984). *Scientific explanation and the causal structure of the world*. New Jersey: Princeton University Press.

Savage, L.J. (1954). *The foundations of statistics*. New York: Wiley.

Shannon, C.E.A. (1948). A mathematical theory of communication. *Bell Systems Technical Journal, 27*, 379–423.

Shapere, D. (1982). The concept of observation in science and philosophy. *Philosophy of Science, 49*, 485–525.

Skov, R.B., & Sherman, S.J. (1986). Information-gathering processes: Diagnosticity, hypothesis-confirmatory strategies, and perceived hypothesis confirmation. *Journal of Experimental Social Psychology, 22*, 93–121.

Slowiaczek, L.M., Klayman, J., Sherman, S.J., & Skov, R.B. (1992). Information selection and use in hypothesis testing: What is a good question, and what is a good answer? *Memory and Cognition, 20*, 392–405.

Snyder, M., & Swann, W.B. (1978). Hypothesis-testing processes in social interaction. *Journal of Personality and Social Psychology, 36*, 1202–1212.

Spellman, B.A., López, A., & Smith, E.E. (1999) Hypothesis testing: Strategy selection for generalising versus limiting hypotheses. *Thinking and Reasoning, 5*, 67–91.

Sperber, D., Cara, F., & Girotto, V. (1995). Relevance theory explains the selection task. *Cognition, 57*, 31–95.

Sperber, D., & Wilson, D. (1986). *Relevance: Communication and cognition*. Oxford: Blackwell.

Suppe, F. (1977). *The structure of scientific theories*. Illinois: University of Illinois Press.

Swann, W.B., & Giuliano, T. (1987). Confirmatory search strategies in social interaction: How, when, why, and with what consequences. *Journal of Social and Clinical Psychology, 5*, 511–524.

Toulmin, S. (1961). *Foresight and understanding*. London: Hutchinson.

Trope, Y., & Bassok, M. (1982). Confirmatory and diagnosing strategies in social information gathering. *Journal of Personality and Social Psychology, 43*, 22–34.

Trope, Y., & Liberman, A. (1996). Social hypothesis testing: Cognitive and motivational mechanisms. In E.T Higgins & A.W. Kruglanski (Eds.), *Social psychology: Handbook of basic principles* (pp. 239–270). New York: Guilford Press.

Tschirgi, J.E. (1980). Sensible reasoning: A hypothesis about hypotheses. *Child Development*, *51*, 1–10.

Tukey, D.D. (1986). A philosophical and empirical analysis of subjects' modes of inquiry in Wason's 2–4–6 task. *Quarterly Journal of Experimental Psychology Psychology*, *38*, 5–33.

Tversky, A., & Kahneman, D. (Eds.). (1982). *Judgement under uncertainty: Heuristics and biases*. Cambridge: Cambridge University Press.

Tweney, R.D., Doherty, M.E., & Mynatt, C.R. (1981). Hypothesis testing: The role of confirmation. In R.D. Tweney, M.E. Doherty, & C.R. Mynatt (Eds.), *On scientific thinking* (pp. 115–128). New York: Columbia University Press.

Tweney, R.D., Doherty, M.E., Worner, W.J., Pliske, D.B., Mynatt, C.R., Gross, K.A., & Arkkelin, D.L. (1980). Stategies of rule discovery in an inference task. *Quarterly Journal of Experimental Psychology*, *32*, 109–123.

Vallée-Tourangeau, F., Austin, N.G., & Rankin, S. (1995). Inducing a rule in Wason's 2–4–6 task: A test of the information-quantity and goal-complementarity hypotheses. *Quarterly Journal of Experimental Psychology*, *48A*, 895–914.

van Benthem, J. (1989). Logical semantics. In H. Schnelle & N.O. Bernsen (Eds.), *Logic and linguistics* (pp. 109–127). Hove, UK: Lawrence Erlbaum Associates Ltd.

Van Duyne, P.C. (1974). Realism and linguistic complexity. *British Journal of Psychology*, *65*, 59–67.

Van Duyne, P.C. (1976). Necessity and contingency in reasoning. *Acta Psychologica*, *40*, 85–101.

Van Wallendael, L.R. (1995). Implicit diagnosticity in an information buying task: How do we use the information that we bring with us to a problem? *Journal of Behavioural Decision Making*, *8*, 245–264.

Wagenaar, W.A., van Koppen, P.J., & Crombag, H.F. (1993). *Anchored narratives: The psychology of criminal evidence*. London: Harvester Wheatsheaf.

Wason, P.C. (1960). On failure to eliminate hypotheses in a conceptual task. *Quarterly Journal of Experimental Psychology*, *12*, 129–140.

Wason, P.C. (1966). Reasoning. In B.M. Foss (Ed.), *New horizons in psychology I* (pp. 135–151). Harmondsworth, UK: Penguin.

Wason, P.C. (1968). Reasoning about a rule. *Quarterly Journal of Experimental Psychology*, *20*, 273–281.

Wason, P.C. (1969). Regression in reasoning? *British Journal of Psychology*, *60*, 471–480.

Wason, P.C. (1977). Self-contradictions. In P.N. Johnson-Laird & P.C. Wason (Eds.), *Thinking: Readings in cognitive science*. Cambridge: Cambridge University Press.

Wason, P.C., & Evans, J.St.B.T. (1975). Dual process in reasoning? *Cognition*, *3*, 141–154.

Wason, P.C., & Johnson-Laird, P.N. (1970). A conflict between selecting and evaluating information in an inferential task. *British Journal of Psychology*, *61*, 509–515.

Wason, P.C., & Johnson-Laird, P.N. (1972). *Psychology of reasoning: Structure and content*. London: Batsford.

Wason, P.C., & Shapiro, D. (1971). Natural and contrived experience in a reasoning problem. *Quarterly Journal of Experimental Psychology*, *23*, 63–71.

Wetherick, N.E. (1962). Eliminative and enumerative behaviour in a conceptual task. *Quarterly Journal of Experimental Psychology*, *14*, 246–249.

Wharton, C.M., Cheng, P.W., & Wickens, T.D. (1993). Hypothesis-testing strategies: Why two goals are better than one. 32nd Annual Meeting of the Psychonomic Society (1991, San Francisco, California). *Quarterly Journal of Experimental Psychology*, *46A*, 743–758.

Yachanin, S.A., & Tweney, R.D. (1982). The effect of thematic content on cognitive strategies in the four-card selection task. *Bulletin of the Psychonomic Society*, *19*, 87–90.

Zuckerman, M., Knee, C.R., Hodgins, H.S., & Miyake, K. (1995). Hypothesis confirmation: The joint effect of positive test strategy and acquiescence response set. *Journal of Personality and Social Psychology*, *68*, 52–60.

Appendices

APPENDIX 1

x_1 is the forecast "Stormy Weather". The likelihood ratio is .90/.30.
x_2 is the forecast "Calm Weather". The likelihood ratio is $(1 - .90)/(1 - .30) = .10/.70$.
H_1 is the hypothesis that it will actually be stormy. $p(H_1) = .50$.
H_2 is the hypothesis that it will actually be calm. $p(H_2) = .50$.
According to Bayes' theorem:

$$\frac{p(H_1 \mid x)}{p(H_2 \mid x)} = \frac{p(x \mid H_1)}{p(x \mid H_2)} \cdot \frac{p(H_1)}{p(H_2)}$$

Thus:

$$\frac{p(H_1 \mid x_1)}{p(H_2 \mid x_1)} = \frac{.90}{.30} \cdot \frac{.50}{.50} = \frac{.45}{.15} = \frac{3}{1}$$

and:

$$\frac{p(H_1 \mid x_2)}{p(H_2 \mid x_2)} = \frac{.10}{.70} \cdot \frac{.50}{.50} = \frac{.05}{.35} = \frac{1}{7}$$

After observing x_1, H_1 becomes three times more probable than H_2.
After observing x_2, H_2 becomes seven times more probable than H_1.

APPENDIX 2

When turning the not-Q card, two outcomes are possible: Either it is an element of the set P, which has N(P) elements, or an element of the set not-P (of size N(not-P)). If the number of not-Q cards in the P set is k, then the probability that the displayed not-Q card is issued from the P set is the proportion of this number k to the total number of not-Q cards in both sets P and not-P:

$$\frac{k}{k + N(\text{not-Q in not-P})} \tag{1}$$

To further calculate this probability, the subject needs to know how many not-Q cards are in the not-P set. This is not given in Kirby's (1994) stimulus materials. We assume that there are 50% not-Q cards in the not-P set. There are 100 cards printed. Depending on condition, N(P) are printed with a P, and 100 – N(P) with a not-P. This is known to the participant. Thus, the probability of finding a P with the not-Q card can further be calculated:

$$\frac{k}{k + [(100 - N(P)) \cdot .50]} \tag{2}$$

However, the participant does not know the value of k. He or she does not know how many not-Q cards there are in the set P. In other words the participant does not know how many printing errors have been made. But, since it is known that the computer prints 1 not-Q in every 10 prints, on average, a probability distribution of the possible values of k can be calculated.

$$p[N(\text{not-Q in P}) = k] = \binom{N(P)}{k} \cdot \left(\frac{1}{10}\right)^{k} \cdot \left(\frac{9}{10}\right)^{N(P)-k} \tag{3}$$

Now, we can calculate further the probability of finding a P on the back of the not-Q card, by considering all possible values of k, and their probability. This can be done by multiplying the probability distribution of k (3) with the probability of finding a P with a not-Q card (2), assuming that the number of not-Q cards in the set P is indeed k.

$$p(P \mid \text{not-Q}) = \sum_{k}^{N(P)} p[N(\text{not-Q in P}) = k] \cdot \frac{k}{k + [(100 - N(P)) \cdot .50]} \tag{4}$$

As can be seen in equation 4, the probability of a P with the not-Q card indeed increases, as the number of P (N(P)) increases: Indeed, when the number of P is highest, the number of not-P's is zero. Hence, the probability of a P with the not-Q card gets its highest value 1.

Author index

Subject index

Note: page numbers in *italics* refer to figures, page numbers in **bold** refer to tables.